VOID

Library of
Davidson College

Telematics and Government

COMMUNICATION AND INFORMATION SCIENCE

A series of monographs, treatises, and texts

Edited by

MELVIN J. VOIGT

University of California, San Diego

WILLIAM C. ADAMS • Television Coverage of the Middle East
HEWITT D. CRANE • The New Social Marketplace: Notes on Effecting Social Change in America's Third Century
RHONDA J. CRANE • The Politics of International Standards: France and the Color TV War
HERBERT S. DORDICK, HELEN G. BRADLEY, AND BURT NANUS • The Emerging Network Marketplace
GLEN FISHER • American Communication in a Global Society
EDMUND GLENN • Man and Mankind: Conflict and Communication Between Cultures
BRADLEY S. GREENBERG • Life on Television: Content Analyses of U.S. TV Drama
JOHN S. LAWRENCE AND BERNARD M. TIMBERG • Fair Use and Free Inquiry: Copyright Law and the New Media
ROBERT G. MEADOW • Politics as Communication
WILLIAM H. MELODY, LIORA R. SALTER, AND PAUL HEYER • Culture, Communication, and Dependency: The Tradition of H. A. Innis
VINCENT MOSCO • Broadcasting in the United States: Innovative Challenge and Organizational Control
KAARLE NORDENSTRENG AND HERBERT I. SCHILLER • National Sovereignty and International Communication: A Reader
DALLAS W. SMYTHE • Dependency Road: Communications, Capitalism, Consciousness and Canada
HERBERT I. SCHILLER • Who Knows: Information in the Age of the Fortune 500

In Preparation:

WILLIAM C. ADAMS • Media Coverage of the 1980 Campaign
WILLIAM C. ADAMS • Television Coverage of International Affairs
MARY B. CASSATA AND THOMAS SKILL • Life on Daytime Television
ITHIEL DE SOLA POOL • A Retrospective Technology Assessment of the Telephone
OSCAR H. GANDY, JR. • Beyond Agenda Setting: Informtion Subsidies and Public Policy
BRADLEY S. GREENBERG • Mexican Americans and the Mass Media
CEES J. HAMELINK • Finance and Information: A Study of Converging Interests
ROBERT M. LANDAU • Emerging Office Systems
VINCENT MOSCO • Pushbutton Fantasies
KAARLE NORDENSTRENG • The Mass Media Declaration of UNESCO
JORGE A. SCHNITMAN • Dependency and Development in the Latin American Film Industries
INDU B. SINGH • Telecommunications in the Year 2000: National and International Perspectives
JENNIFER D. SLACK • Communication Technologies and Society: Conceptions of Causality and the Politics of Technological Intervention
JANET WASKO • Movies and Money: Financing the American Film Industry
OSMO WIIO • Information and Communication Systems

Editorial Board: Robert B. Arundale, University of Alaska, Walter S. Baer, Times-Mirror, Jörg Becker, Philipps-Universität Marburg, Erwin B. Bettinghaus, Michigan State University, Brenda Dervin, University of Washington, Nicholas R. Garnham, Polytechnic of Central London, George Gerbner, University of Pennsylvania, James D. Halloran, University of Leicester, Brigitte L. Kenney, Infocon, Inc., Manfred Kochen, University of Michigan, Robert G. Meadow, University of California, San Diego, Vincent Mosco, Temple University, Kaarle Nordenstreng, University of Tampere, Ithiel de Sola Pool, Massachusetts Institute of Technology, Dieter Prokop, Frankfurt, Germany, Everett M. Rogers, Stanford University, Herbert Schiller, University of California, San Diego, Russell Shank, University of California, Los Angeles, Alfred G. Smith, Unviersity of Texas, Austin, Frederick Williams, University of Southern California.

Telematics and Government

Dan Schiller

Ablex Publishing Corporation
Norwood New Jersey 07648

Copyright ©1982 by Ablex Publishing Corporation.

All rights reserved. No part of this book may be reproduced
in any form, by photostat, microfilm, retrieval system, or any other means,
without the prior written permission of the publisher.

Printed in the United States of America.

Library of Congress Cataloging in Publication Data

   Schiller, Dan.
   Telematics and government.

   (Communication and information science)
   Bibliography: p.
   Includes index.
   1. Telecommunication policy—United States.
I. Title. II. Series.
HE7781.S34   384'.068   82-1795
ISBN 0-89391-106-2   AACR2

ABLEX Publishing Corporation
355 Chestnut Street
Norwood, New Jersey 07648

To Anita and Herb

TABLE OF CONTENTS

Acknowledgments   ix
List of Tables   xi
Foreword   xii

Part One
The Course of Postwar Policy for U.S. Telematics   1

*Microwave Radio and Expansion of Private
  Telecommunications   8
Private Terminal Equipment and
  the Public Network: The Carterfone   15
The First Computer Inquiry   22
The Specialized Common Carrier Proceeding   41
Satellites and Value-Added Networks   48
Business Users and the
  Legislative Battle for Position   54
The Emerging Resale Environment   70*

Part Two
The U.S. Offensive In International Telematics   97

*Policy for Integration and Expansion of
  International Telematics   105
The United States Regulatory and Policy Offensive   149*

Part Three
Telematics and Government   189

*The Government Telematics Market   191
"Paperwork Reduction" and Information Management:
  Government as Telematics User   203
Monopoly and the Mail: Government as
  Prospective Telematics Supplier   210*
Bibliography   217
Index   229

ACKNOWLEDGMENTS

It is proper to begin by acknowledging the undoubted favorable influence exercised upon this work by the lovely place in which it came to be written, and the efforts of Sunie Davis to bring us there, to Swans Island.

I should like to thank Robert Lewis Shayon for many animated discussions which proved always to be a provocative and clarifying influence. Chris and Ellen Sterling supplied a warm and hospitable field environment for my forays to Washington, where FCC dockets are kept. Chris Sterling, Anthony Rutkowski and Dennis L. Rodkin sent materials which otherwise might not have found me. My emphasis on the policymaking role of telematics users and user groups returns ultimately to an essay by Tim Haight.

Through several reincarnations my classes on communications technology at Temple University have been of signal assistance in helping me to try out arguments and at times, even better, to stick to my guns; I wish therefore to thank my students in those seminars.

Though few, readers of this book have so far been avid—and vastly helpful through their criticism. I should like to acknowledge Sunie Davis, Vinny Mosco, Marcus Rediker, Anita, Herb and Zach Schiller and Dallas Smythe for carrying out this task.

Dave and Mary Ann Davis gave me the run of their kitchen and permitted me to inflict a number of peppery dishes on them during welcome periods of respite from writing.

Finally I must note that such acuity as this work may possess owes in large part to the untiring persistence of vision of two scrupulous and imaginative teachers, to whom I dedicate it—Anita R. Schiller and Herbert I. Schiller.

List of Tables

TABLE 1	Land/Fixed Stations in the Industrial Radio Services 1950–1975	14
TABLE 2	Growth of Business-Oriented Services Offered by the Bell System	15
TABLE 3	Industrial Radio Services Portable Mobile Units 1950–1974	16
TABLE 4	Estimated Distribution of Computer Installations by User Industry	24
TABLE 5	Top 20 Processing Services Firms	81
TABLE 6	Overseas Telephone Services—Number of Calls	101
TABLE 7	U.S. International Telex Traffic: Top 25 Correspondents 1979	102
TABLE 8	The World's Submarine Cables	105
TABLE 9	Telephone Construction Expenditures 1979—Top 25 Nations	106
TABLE 10	Number and Value of Computers and Related Equipment in Use	106
TABLE 11	Trade Barriers to Telecommunications, Data and Information Services	124
TABLE 12	Top 100 Data Processing Companies and Top 100 DoD Contractors 1979 Overlap	193
TABLE 13	Top Eight DP Companies and the Government Computer Market 1979	197
TABLE 14	1979 IBM Market Share of Selected Government Computer Submarkets	200
TABLE 15	Rank-Order of Computer Manufacturers by Number of Computers Per Agency	201
TABLE 16	Contract Services as a Percentage of Total ADP Operating and Capital Costs by Government Agency 1979	202

FOREWORD

A decade ago one rarely heard about "information technology," to say nothing of "telematics," despite the rapid convergence of telecommunications and computerized data processing into a unitary technology, market, and concept that is the reality these terms capture and express.* Today, in contrast, references to a new "electronic cottage" litter popular discourse; "high technology" evokes elegance; an "information age" appears by media magic on the societal horizon. Cosmic explanations declare the waning of older social arrangements in favor of post-industrial society, and a virtual welter of video displays and home-computer terminals.

Talk is cheap. The momentous implications of an increasingly computerized society *do,* nonetheless, demand the most thorough and exacting historical scrutiny. This book contributes to a discussion that, though barely beginning, must find an intensive and open-ended development.

Computerization of society has been no random or simple market-driven process. Not only has the Federal Government supported telematics ceaselessly through huge contracts for advanced computer communications, but for thirty years telematics has been driven forward steadily through deliberate Government policy decisions.

The role of Government in developing telematics, paradoxically, has been poorly understood, because the critical part played by telecommunications in *private* economic growth historically has been generally neglected. Communications have not been treated seriously enough in appraising the progress of the American—or, for that matter, any other—economy during the Industrial Revolution. Recent work by Allan Pred (1973, 1980), Thomas Cochran (1977) Alfred Chandler (1977), Eric Hobsbawm (1975) and Richard Du Boff (1980), however, has begun to underscore the profound importance of the telegraph to business growth.

According to Du Boff, the telegraph reduced corporate uncertainty and risk as it allowed markets to be unified into an increasingly cohesive national and international system (Ibid.). Pred (Ibid.) finds that the speed-up of business that telegraphy facilitated not only supplied an incentive to increase production but enhanced control over shipments and inventories. No less than railroad transport, electricity or machine tools, the telegraph enabled development of geographically dispersed business enterprises. Telecommunications proved historically to be critical to effective corporate coordination and control over multi-unit operations.

---

*The word is borrowed from the French "télématique," a neologism coined by Nora and Minc (1980).

To the decisive contributions made by telegraphs and telephones (the latter at first employed for local communications), an unending set of added benefits has been conferred on business through application of computer communications to productive, distributional and administrative functions. The utilizable speed afforded by telegraphy and telephony was limited always by the substantive information-processing capabilities of business users; the more quickly orders piled up as a result of reliable telecommunications, for instance, the greater became the need to find means of processing them—of keeping up. Progressive advance in processing of data, first by manual and electro-mechanical punch-card techniques, then by electronic digital computers, became a business imperative.

Computers and telecommunications began to converge—a process expedited by invention of transistors in the late 1940s and integrated circuits a decade later. Innumerable business applications of computer communications began to be explored. At this stage, the role of the U.S. Government became crucial.

This was not only because tax monies continued to be used in support of ever more powerful and sophisticated computer communications applications. Telegraphy and telephony are regulated under the Communications Act of 1934 by the Federal Communications Commission (they also are subject to regulation by state public utility commissions). Subject to extensive exit, entry and rate regulation by the FCC, telecommunications simply could not be married felicitously to the computer without the Government's consent. And, far from being a one-time thing, regulatory approval needed to be extended to countless unfolding technical, economic and policy issues. Most important, regulatory acquiescence to the progressive merger of telecommunications and computers was contingent on settlement of a widening and deepening dispute.

American Telephone and Telegraph, the historical core of this country's telecommunications system, with 1,500-odd independent telephone carriers, wished to be certain that *its* network would be harnessed to computers *if and when it found it profitable and expedient*. Unregulated computer companies, in contrast, sought a contrary course more favorable to *their* own corporate strategies for growth.

Of itself such a contest might appear a standoff. Yet the relevant regulatory agency, the Federal Communications Commission, has—we shall see—consistently and steadfastly overruled the immediate interests of the regulated telephone companies through seven administrations spanning a quarter of a century. What accounts for this decisive policy stance?

The Commission in fact responded to the one and only economic entity able to muster greater collective force than AT&T—that is, to big business as a whole, in the shape of present and prospective *users* of advanced computer communications. Business users, it is the major burden of this book to point out, have become a pivotal and still-growing telematics policymaking force. This, despite having been entirely overlooked by analysts seeking to explain the historical course of policy in computer communications, but who have neglected business demand for integrated telematics applications. Geographically-dispersed big businesses need computer communications today not one whit less than their forbears needed the telegraph in 1850! If, at times, I overemphasize the cardinal role of such business users below, this is because that role has virtually never up till now been probed.

The work is divided into three parts. Part One traces the evolution of postwar policy for domestic telematics—in parallel with growing corporate demand for merged computer-communications services put to private use. Part Two extends the argument to the international sphere, because the structure of corporate enterprise is now essentially transnational. Part Three returns to Government's other critical role in the computerization of society, as a market for advanced telematics equipment and services. As we shall find, as a user itself, Government again continues to further the progress of telematics under corporate guidance. It has not therefore been some inert and mute new technology called telematics that has carried us over the threshold of an information age. The reality is that business users demanding advanced telematics services have mustered policymakers' support effectually, so as to enhance their private control over not merely information technology—but our economy and society as a whole.

Telematics
and
Government

ONE

The Course of Postwar Policy for U.S. Telematics

"What is the basic aim of this country's telecommunications policy? Is it. . .to promote the widest availability of high quality communications service to all of the people of the United States, or does that aim now yield to particularized interests of specialized classes of users. . . ."
——— John D. deButts, Chairman, AT&T, 1976

"Down through the years, the telephone companies have priced their services to support the goal of universal service. No longer is this policy tenable. More and more we must relate rates to cost - service by service, customer by customer."
——— Charles L. Brown, Chairman, AT&T, in Tanenbaum, 1982

"Lest we forget, the idea of a 'haves and have-nots' society/marketplace could become a reality right along with rooftop reception dishes."
——— Advertising Age, 1981

Telematics - and telematics policy - today reach directly into the heartland of the business system. Innumerable applications of merged computer-communications permit unprecedentedly rapid, flexible and inclusive control over diverse managerial and administrative and productive and distributional functions. Both nationally and internationally telematics has become the key to development of new services, whose heavy information content is inseparable from the equipment and software employed in their conveyance and distribution.

Only a long string of Government regulatory and policy decisions enabled telematics to realize this technical and economic potential. Only deliberate policy choices made feasible the unparalleled centralization of business control that telematics engenders. Yet policymakers, far from acting on their own initiative, have tended to assume a *reactive* role in expediting integration of telematics into the business system. The very largest prospective corporate *users* of telematics, banding together gradually into trade associations and other industry groupings for this special purpose, have effectively *forced* policymakers to respond to their increasingly insistent demands for efficient computer-communications.

Business users had to press themselves upon regulators simply because the regulated telecommunications industry itself was not meeting their dynamic telematics needs. Both the terms and the costs of use for developing equipment and services often set the traditional telecommunications industry and the largest business users at odds. As a result of this spiraling conflict, the entire shape of the traditional telecommunications industry has been transformed and, in our era, business users of telematics have become a decisive policymaking force.

I shall survey below a more or less standard litany of "procompetitive" decisions by the Federal Communications Commission—the key locus for policymaking in this field—to show that telematics-user influence has been neither slight nor temporary, but rather central to the evolving shape of policy as a whole.

We must dwell briefly, first, on the historical context in which telematics developed, for this development occurred against a background of phenomenal economic and corporate expansion. Several ready indices pinpoint relevant trends. Gross U.S. energy consumption, underwriting the entire process of business growth, increased from 33.6 to 78 quadrillion British Thermal Units between 1950 and 1979 (U.S. Department of Commerce, 1981: 110). Reliance upon coal dropped dramatically (from 38 percent of total consumption in 1950 to 19 percent in 1979 (Ibid.: 111)); oil and natural gas increased their share from 57.5 to 73.1 percent (Ibid.). Petroleum linkages to chemicals, plastics and rubber substitutes grew firmer; in the transport sector, the growth of automobile and airline industries cemented dependence upon petroleum. Internationally, analogous increases in demand for energy occurred throughout the industrial West.

Enlarging demand has inspired constant intensification in processes of exploration and exploitation of energy resources. In turn we find drilling at ever more remote locations, including offshore at many miles distance and at equally inhospitable terrestrial sites (such as Alaska's North Slope). Pipelines and refineries, built in concert with this intensified search for petroleum and natural gas, encircle the globe with chains of dependence upon non-renewable energy. *Telecommunications* furnishes the vital coordinating link between far-flung energy exploration and production and distribution operations. And, coupled with data processing, it is telecommunications that permits coordination and control of dispersed drilling, refining and transport functions, from satellite exploration to final drill site and from wellhead to ultimate consumer.

One or two links down the economic chain, telecommunications now plays an equally impressive role in manufacturing, merchandising and retailing. A gross index to this trend is glimpsed in nonresidential fixed investment in structures and producers durable equipment—which rose from $19.2 billion in 1945 to $163.3 billion in 1979 in constant 1972 dollars (*Economic Report Of The President*, 1981: 234). As construction and fixed investment boomed, new manufacturing plants and mills and wholesale houses and retail stores were built. Simultaneously, the number of business telephones increased from 8.7 million in 1945 to 46 million in 1979 (Federal Communications Commission, *Annual Report 1946*: 26; Federal Communications Commission *Statistics Of Communications Common Carriers* 1979: 7). As we shall find, only telematics could control and unify the complex industrial and commercial operations thereby engendered under centralized corporate command.

Telecommunications is as important to routine linkages between office buildings as between industrial plants and warehouses. The postwar economic expansion was financed in part by banks; measured in current dollars, commercial bank loans and investments increased from $113 billion in 1948 to $1,132.5 billion in 1979 (*Economic Reort Of The President 1981*: 303). Persuaded first to make use of banks for savings and checking, and then to come to depend upon them as well as other financial services organizations for credit to underwrite individual purchases, consumers flocked to banks by droves through the postwar years. In 1946 roughly 14,000 banks operated but 4,000 branches. With liberalization of branching laws, by 1979, 14,688 banks operated fully 39,308 branches within the United States (U.S. Department Of Commerce, 1981: 502). Savings accounts totaled $6.3 billion in 1946 and, after some years of slow growth, shot up to well over $100 billion by the late 1970s, again in current dollars (*Economic Report Of The President 1981:* 261). Outstanding consumer credit was enlarged unceasingly, from $25.6 billion in 1950, to $65.1 billion in 1960 during a decade of low inflation, to $143.1 billion in 1970 and $380.5 billion in 1979, still in current dollars (Ibid.: 310). Fueled not only by automobile purchases but also by the myriad other consumer expenditures evidently required by modern households, the individuated complexity of consumer credit itself necessitates reliance upon increasingly sophisticated telematics services and equipment.

The telephone network, acting as a critical segment of the economic infrastructure, also expanded mightily. The number of business and residential instruments supplied to service subscribers of telephone carriers filing reports with the FCC grew from 28.3 million in 1946 to 153.1 million in 1979; the number of local calls originating from these telephones rose from 48.1 billion in the former year to 268.7 billion in the latter. Toll calls, amounting to a few over 2 billion in 1946, reached almost 4.2 billion in 1961 and then exploded to nearly 29.9 billion by 1979 (Federal Communications Commission, *Statistics Of Communications Common Carriers 1979:* 18). Accelerated demand for voice telephony was matched by steady decline in the cost of long-distance services (Oettinger 1980: 191–192). The Bell System—the core of the domestic public switched network—enjoyed a profitable monopoly over the 80 percent of all telephones which it owned. Protected no longer mainly by patent control and unalloyed market power, but instead by federal regulation itself, AT&T dominated U.S. telecommunications (Brock, 1981: 177–178). Paradoxically this would come to mean that AT&T might be more vulnerable to regulatory decisions and, by extension, to the forces that influenced such decisions, than to telecommunications industry market forces.

The industry as a whole, however, was undergoing decisive change through technical advances—from within and outside it. First, incremental increases in bandwidth (information-carrying capacity) utilizable by long-haul and local telecommunications links have been attained—in coaxial cable, in microwave radio, satellites, fiber optics and cellular radio. More efficient use of available bandwidth through multiplexing and other techniques has been the counterpart to this same tendency. Second, qualitative improvements in the amount of computer processing power available in a given space and at a given price by means of transistors, integrated circuits, microprocessors, large-scale integration and, today, very large-scale integration, have been readily incorporated into a wide range of telematics equipment.

First developed during World War Two to solve time-consuming ballistics problems of pressing relevance to the anti-fascist war effort (Goldstine, 1972), electronic digital computers quickly moved from public to private hands. In 1956 the value of installed computers in the U.S. approached $270 million, as IBM overturned Remington Rand's earlier supremacy to achieve a 75 percent market share (Brock, 1975: 13). In the late 1950s solid state computers making use of cheap transistors were introduced by Univac, IBM, NCR and Honeywell; prices declined while performance was vastly enhanced. By 1959 a billion dollars of installed computer equipment was in place, and this figure rose to $6 billion by 1965. By this time, third generation equipment built around integrated circuits, further improving the price/performance ratio, was introduced (Ibid.: 15–16). The prodigious growth rate of the computer industry has continued unabated since that time. In 1980 the installed base of computers in the United States was valued at about $89 billion, out of a world total of some $160 billion (*EDP Industry Report* 28 May 1980: 12).

Greater bandwidth and cheaper, more powerful computing have fostered a trend toward distributed data processing—by which computer processing capacity is distributed through telecommunications circuits to the locations at which it may be required. Unprecedented efficiencies may result from distributed data processing; more to the point, it portends a more or less total entanglement or intermixture of computing and telecommunications. Such a merger is speeded as well by the equally thorough incorporation of computer and computer-like instrumentation to switch or control allocation of telecommunications circuits for AT&T and other carriers.

The very coherence and smoothness of telematics' technical evolution has nurtured a most far-ranging, complicated and painful regulatory conundrum. Such an impact, indeed, could hardly have been avoided. Telecommunications has long been a regulated industry, subject to exit, entry and price controls; computers by and large have not. The place of regulation in telematics has therefore itself comprised a prominent agenda item over the past twenty-odd years, with the telecommunications industry seeking to make use of its traditional hold over the regulatory process to foster a maximally favorable environment for its full-scale entry into telematics, while the computer industry moves toward this same goal hoping to continue to evade the regulatory net. If this alone were not sufficient to embroil telematics in policymaking dilemmas, there are also numerous junctures at which discrepant engineering and technical standards common to telecommunications and computers respectively must confront and bow to regulatory fiat. Voice telephony places one set of constraints upon the public switched telephone network; computer communications, to the degree it seeks to depend upon the public network, quite another. How shall the pace at which the public network changes to accommodate computers, while retaining its ability to transmit voice traffic, be set, and how shall this monumental transformation be paid for?

These questions, important though they may be, have assumed an urgent nature owing mainly to the very ubiquity of computerized data processing within an expanding business environment. Geographically dispersed corporations, as mentioned, ofter prefer to centralize computer resources and distribute processing power as needed through telecommunications circuits, rather than to invest in redundant computer facilities at diverse sites. Such businesses, again, may prefer to own and operate their own telecommunications links, or at least to lease them at low and stable rates, than to be made subject to the vagaries of monopolistic pricing by a quasi-public telephone company driven both by surging demand for conventional voice circuits and by pricing policies which have not always externalized the costs borne by the largest users to the maximum possible extent. These concerns must also ultimately be viewed in a context of dynamic corporate strategy, both to maximize profitable growth and to brace against competitive decline. When investment plans call for maximal use of cheap labor here, or nearby raw materials there, or convenient access to markets at a third place, telematics is today a basic planning factor. How much will it cost to obtain circuits capable of efficiently transmitting necessary data to a new plant or warehouse or office building at a particular proposed location? Are the requisite circuits even available? How may current telematics expenditures be trimmed? Will improved equipment turn the trick? If so, is it available, and on what terms? Because an ever-widening sphere of U.S. business relies upon telematics, an ever-growing number of businesses are asking—and demanding answers to—these and related questions.

Business began to seek solutions to such problems even before the Second World War was over. IBM and General Electric obtained experimental licenses from the Federal Communications Commission for a Schenectady-New York-Philadelphia-Washington private microwave system for business data transmission in November of 1944 (Brock, 1981: 182). Even as computers, with their unparalleled capacities for keeping track of a lengthening list of business activities, penetrated U.S. industry, telecommunications issues and problems moved rapidly up on the agenda of crucial regulatory concerns. Prompted by large corporate users, regulators began to address the terms and conditions under which communications and, implicitly, computers, were to be made available. Because many critical changes in this area compelled the regulators' acquiescence, the latter were drawn more and more into the center of decisions whose immediate and ultimate impacts would be felt far, far outside the telecommunications industry itself.

The full story of user intervention in telecommunications policy will eventually require analysis not only of hundreds of separate and, taken alone, frequently inconsequential decisions, but also of a variety of private sector policymaking activity having to do with investments, equipment contracts and so forth. Below I can dwell only upon the most visible and, I believe, critical of these proceedings. The series of Federal Communications Commission dockets upon which I have chosen to focus has routinely been hailed as awakening "competition" in the U.S. telecommunications industry (Martin, 1977: 351)—as if competition is some sort of natural economic condition that does not require explication. In this, the late twentieth century, when market concentration in mature markets often means that the dominant "competitor" commands a market share ranging from 30 to 60 or 70 percent, any moves toward real competition necessarily demand explanation. They are in no way natural or transparent. What competition has been engendered in telecommunications, an industry in which the dominant firm has long controlled fully 80 percent of the market, fairly begs for analysis. What competition has arisen in telecommunications—and this is no small issue in its own right—has rightly been called the user's "ultimate insurance policy" (Karten, 1980: 24) and, like all insurance, must be paid for by the user himself. Let us now turn, therefore, to the thrust and parry of the regulatory arena—in which the explicit intervention of the telematics user community has been a notable, if largely unremarked, feature over the course of the past quarter-century.

MICROWAVE RADIO AND EXPANSION OF PRIVATE TELECOMMUNICATIONS

Developed before and during the Second World War, microwave radio emerged relatively unencumbered of patent controls because much of the research and development out of which it grew was performed for the U.S. military. Characteristic of the frequency bands utilized by microwave are the short wavelengths which may be readily focused and transmitted in stright lines for telephone, video or data carriage. Thus the economic superiority of microwave derives from its break with expensive right-of-way privileges required by cable and telephone wire; microwave makes use of relay towers at rough intervals of twenty to thirty miles (Brock, 1981: 180–181).

However, microwave does require frequency assignments to be made by the Federal Communications Commission. In 1948 the FCC reserved permanent use of most microwave frequencies to common carriers—the regulated telephone companies (Ibid.: 183). Because of an assumed shortage of available frequencies in the microwave bands, the Commission in the main continued to withhold permission to construct microwave networks from all but the carriers, television broadcasters unable to obtain necessary service from the carriers, and right-of-way companies—pipelines and railroads. Even so, microwave found increasingly extensive use, both within the Bell system and by private utilities and petroleum companies. By the later 1950s, microwave comprised a substantial portion of telephone and television capacity in the Bell System (22 percent and 78 percent respectively); more important, private systems embraced 14,000 route miles for utilities and 17,000 miles along petroleum pipelines (Brock, 1981: 203). Other prospective applicants were authorized to use microwave by the FCC on a case-by-case basis.

At the end of 1956 the Federal Communications Commission was nudged into reviewing its policy for microwave frequency allocation to permit private users to contruct systems for their own use even where common carrier facilities were already available. In the summer of 1959, after some twenty volumes worth of testimony, the Commission ruled in its "Above 890" decision that microwave should indeed be made available to private users on these more lenient terms. Restrictions on interconnection with AT&T, and on sharing of microwave links by different companies, however, clearly limited use to those large companies with bulk communications needs between two distinct points (Brock, 1981: 206). For these, the very largest users, private microwave furnished an apparently economical alternative to leased lines from the Bell system (Ibid.).

It can hardly be adequate, however, to explain the Above 890 decision (named after the frequencies in the megahertz band whose use was broadened to include private firms) simply as a result of "technological progress in microwave technology and the increasing use of microwave for television and voice transmission" (Ibid.: 203). That greater intensity of use had become technically feasible was not enough, nor was presence of an abstract demand for use of microwave frequencies. Rather, given the zealous attempts by AT&T to protect its control over the nation's telecommunications by appeal to regulatory authorities—attempts made all the more pressing by a 1956 Consent Decree which forced Bell to make its patents more widely available and therefore bit into a previous paramount source of its market power—private microwave had to be actively fought for by prospective business users.

Actually, the Above 890 hearings established an enduring precedent in the form of a powerful alliance between independent equipment manufacturers of electronics and computer instrumentation and potential users of private systems. The Electronic Industries Association furnished a lengthy engineering study, for example, in support of the claim that many more microwave frequencies might be used without interference—and that EIA affiliated manufacturers *themselves* had a vital need for private microwave circuits. By opening up microwave to private use, of course, independent equipment manufacturers stood to benefit as producers of the facilities required by companies intent on innovating private networks (FCC, "In The Matter Of Allocation Of Frequencies in the Band Above 890 Mc," Docket Number 11866, 4837–4984.)

Of course, the common carriers resisted any other outcome than that delivery of microwave signals should be confined to their own networks (for comments of the Bell System see Ibid.: 2017–2332). AT&T was resisted, however, by a countervailing force far more powerful than the electronics industry taken alone. It is first in the Above 890 proceeding that the prospective *users* of an incipient telecommunications offering gathered to oppose Bell in at least a loosely organized fashion.

The American Newspaper Publishers Association wrote to state that "private communications users should have a full right to participate in the advantages of microwave" (Ibid.: 19 July 1956, 39). "(W)e feel," ANPA's General Manager Cranston Williams declared, "the frequency space should be allocated looking towards future expansion of communications for newspaper use."

J. E. Caldwell, Controller of Montgomery Ward, amplified on the nature of the need felt for private microwave circuits by his company (Ibid.: 4 February 1957). Montgomery Ward operated mail order houses in nine cities, he noted, retail outlets in 562 locations and catalog stores at an additional 425, spread throughout the United States. "Operations are controlled both from our Chicago office and from regional offices and mail order houses." Caldwell believed that in light of this geographically dispersed network of chain outlets, "(e)conomical and efficient communications methods are of prime importance to our company."

For just this reason Ward was "constantly conducting research into improved automatic communications methods and new applications requiring additional communication facilities." Study showed that reliance upon automatic equipment utilizing microwave frequencies was "feasible and would provide better and more economical transmission of our traffic." As a heavy user of communications circuits in an expanding economic environment Montgomery Ward was also an innovator in this field, customizing and configuring communications technology to its own particular needs. Such innovation was essential to well-planned growth of the company. Caldwell continued: "(r)esearch in the field of data processing indicates that there will be a future need for prompt and adequate communication for the purpose of transmitting accounting and other data to be used in electronic computers or similar devices. Transmission of customer orders, source orders, and similar transactions would create additional requirements for equipment of this kind."

The National Retail Dry Goods Association, representing some 8,100 stores ranging from small firms to enormous enterprises like Ward and Sears, joined the proceeding as well. Plainly speaking for its larger members primarily, the NRDGA concurred with Caldwell that retailers "should have complete freedom to choose the most desirable method" of communicating from point to point. This freedom of choice, Caldwell summarized, "does not exist under present conditions of intra-state and inter-state franchise monopoly.

> In the past, it has often been impossible to add needed facilities because existing rates were not economical, and no choice of other methods was available. In view of existing and future needs for better and cheaper communications methods, it is the position of Montgomery Ward that the subject frequencies should be so allotted as to make a portion of the band available for commercial purposes.

The company—and its trade association, the NRDGA—sought room for experiment on their own terms (see also Ibid.: 1 July 1957, 2828–2943). How else would it be possible to cater effectively to mushrooming demand, as consumer installment credit exclusive of automobiles shot up from about $9.5 billion outstanding per month to $27 billion between 1950 and 1960 (*Economic Report Of The President 1981*: 310).

Dynamic coalescence of user and innovator roles was by no means the province only of dry goods retailers. In a decade during which the amount of outstanding automobile credit also increased threefold, from $6 to $18 billion (Ibid.), it should not surprise us that the automobile manufacturers were also present in force throughout the Above 890 proceeding. Communications had become a key factor in growth of automobile productive facilities. Its request for a hearing arose, the Automobile Manufacturers Association said (FCC, "In the Matter of Allocation of Frequencies in the Band above 890 Mc," Docket Number 11866,

15 March 1957, 849), "from this industry's usual practice of utilizing all available new methods, equipment, facilities and techniques to improve quality and production, thereby reducing product cost." As the auto industry depended upon "one of the most extensive communications networks in the world," its incessant search for "more efficient and economical methods" had "extended into this field."

In particular the auto industry had found that "(c)ontinued advances in production and accounting methods have resulted in increased requirements for economical and dependable communications systems to handle voice communications, teletype, industrial television, data processing, computer programing material and other data of similarly specialized nature." Experiments had already shown that private microwave transmission of data carried distinct cost advantages in comparison with common carrier charges for private line (leased circuits dedicated to the exclusive use of the subscriber but owned, provided and maintained by the common carrier) (Ibid.: 849–850). Critical, too, microwave would facilitate centralized management control of remote operations both intra- and interplant. "As these facilities would control production, accounting and distributive functions," the automobile group explained, "it is highly desirable that complete control of these facilities and thus the functions which they regulate, be available to management." Private circuits would ensure "maximum output of defense and commercial goods" without interruption both in peacetime and "during times of civil unrest, acts of God, etc." (Ibid.: 850–851). As the auto industry dispersed its manufacturing plants out of Michigan—indeed, out of the midwest—communications steadily assumed more vital relevance to production. Echoing the retailers, the auto companies asserted that they "should have the same latitude in the use and implementation of (their) communications facilities that (they) enjoy in the use and implementation of the many thousands of other tools, facilities and services necessary to the conduct of (their) business." It was not merely costs, but also and especially this ability of management to configure and control communications as required by larger plans for expansion, that dictated the automobile association's position. Heavy use and controlled innovation were tied hand in hand with private corporate strategy.

Elsewhere in the proceeding, Victor G. Reis, Chairman of the National Association of Manufacturers' Committee on Manufacturers Radio Use, took up a similar theme (Ibid.: 27 May 1957, 1585–1596). NAM's intense interest in private microwave, Reis confided, "is based upon our need to use every means available for. . .eliminating inefficient, wasteful or otherwise uneconomical procedures.

> The manufacturing companies need the best means of communications that are available to the same degree that they need the most efficient productive machinery, the most effective procedures and the highest grade of raw materials. (Ibid.: 1585–1586).

Although but few manufacturing firms were actually using point-to-point microwave, the manufacturing community well realized "the possibilities of microwave." "(W)e want to make it clear that we have a high degree of interest in the allocation of frequencies in the bands Above 890 megacycles," Reis told the FCC. Microwave, he explained, offered the fastest known method of communications "capable of handling operational, administrative and informational traffic in large volume at a substantially lower cost, and with a high degree of reliability and a low rate of upkeep" (Ibid.).

Manufacturers had not made greater use of microwave only because of the Commission's restrictive rules—which treated applications for service on a case by case basis and thereby defied prediction with respect to the outcome of any particular application. NAM cited examples of hardy applicants who had had applications pending before the Commission for two years, still with no guarantee of success. In spite of such delays, nonetheless, various manufacturers "maintained a steady interest in microwave developments in the Commission" (Ibid.: 1586). Union Carbide, for instance, sought to link buildings then being erected in Westchester County with its Park Avenue address through between thirty and eighty talking circuits, half a dozen teletype circuits and additional circuits for data processing information. "Similar expressions of interest could be produced from other large industrial organizations," asserted NAM; for national manufacturers had "an important stake in the allocation of microwave frequencies for general private use" (Ibid.: 1587–1588). Another firm's cost studies indicated that not only would private microwave yield substantial cost savings over leased lines but, in addition, "we would own the system rather than having to continuously pay lease charges" (Ibid.: 1588). Presumably ownership allowed depreciation write-offs and conferred other tax advantages as well; equally vital, it lent stability to the planning process.

NAM viewed the possibility of allocation of the frequencies above 890 megahertz to the common carriers exclusively "with the greatest concern, and in fact feel that it is the nub of this whole proceeding" (Ibid.: 1592). Such a decision would amount to an unauthorized gift "of exclusive control of microwave frequencies to a special class of private business," and therefore would install "a pure monopoly of the microwave spectrum by a private corporation" (Ibid.: 1593). The shortcomings of this eventuality were legion. Not only would "complete control" over the disposition and use of microwave bands pass to the common carriers, but this itself would mean that "(t)he technical development of microwave communications and new kinds of applications would be limited to those originating with the common carriers thus eliminating competition and restricting future developments to those resulting in the greatest benefit to the common carriers" (Ibid.: 1593). The manufacturers of the United States would thereby be prevented from innovating to meet their specific needs, present and planned, in the increasingly vital area of communications. Freedom of choice—for them—would be lost (Ibid.: 1596).

Perhaps the clearest statement of interest in the proceeding was that of the Central Committee on Radio Facilities of the American Petroleum Institute. The Petroleum Radio Service, established by the Commission in 1949, had been allotted limited private microwave frequencies on a shared basis with other Safety and Special Radio services (today these categories of utilization of the electromagnetic spectrum fall under the rubric of Private Radio Services). "By virtue of being willing to make full use of this means of communications, even though in the developmental stage, the petroleum industry has contributed to the development of the art" (Ibid.: 16 January 1957, 298). At a cost of many millions, the petroleum industry had installed and was operating hundreds of stations covering thousands of system miles. "In the few short years since 1949," API declared, when the petroleum industry pioneered the first industrial microwave installations, "it has become the standard means for providing the petroleum industry's heavy requirements for point-to-point communications." By the time of the Above 890 docket, microwave facilities found use in such fields as telemetering, industrial television, data processing, supervisory control, alarm signaling and coordination of VHF mobile radio services—all of which had become urgent in themselves, owing to "fast changing petroleum technology" (Ibid.). Communications, once again, had become a critical factor in production; for fast-developing offshore oil operations, for example, microwave furnished "the only means available to meet the industry's communications requirements" (Ibid.: 299), thereby offering "the great flexibility and reliability which are so essential." Assignment of microwave frequencies to the petroleum industry, "without restriction," had become a question of "prime importance . . .and thus is in the highest public interest" (Ibid.: 300). Granting microwave to common carriers exclusively would detract from the public interest, on the other hand, because the carriers would furnish "a *general service* to the public rather than a *specialized* or *custom type* of *service* to an individual industry." The specialized technical needs of prospective users, under such circumstances, would be "foreign" to the carriers' chief responsibility—provision of general purpose service (Ibid.: 300–301). Control and operation of communications by microwave—since the latter constituted "truly a plant facility"—should pass to the oil companies when demanded (Ibid.).

Despite restrictions imposed by the Commission on private microwave users—sharing of facilities was disallowed, while interconnection between private systems and common carrier networks would be determined on a case-by-case basis (Brock, 1981: 205)—there can be little question that the very authorization of private microwave systems was a decisive policy shift. The Above 890 proceeding and its result signified an emerging shift in the balance between public and private networks, and comprised a key early verdict over the terms upon which new communications technology would be integrated as a strategic factor in production.

The decision provoked a competitive response from AT&T, of which more later, in the form of a heavily discounted bulk private line service called Telpak.

Telpak was intended to deter large users, precisely those who benefited from the Above 890 decision, from building their own private systems. By undercutting the costs users would face were they to furnish themselves with the latter, Telpak tariffs frequently found their desired effect and induced business users to migrate to Telpak itself. So tenacious were the property rights which users subsequently established in Telpak over the years that, despite the Commission's finding in 1976 that Telpak was illegal and discriminatory because it was priced below the cost of capital employed in providing it, and despite AT&T's *own* eventual desire to eliminate it, Telpak persisted until mid-1981 (Brock, 1981: 207–209).

In concluding that sufficient microwave frequencies existed to accommodate both carriers and private users the Above 890 ruling decisively opened communications to business users' unfolding demands. By 1975 nearly 600 private microwave systems were in operation (American Enterprise Institute, 1980: 13); since 1959, according to Wiley (1981: 49), some 190,000 miles of private microwave networks have been constructed. Private microwave use, crucially, developed together with increased private exploitation of many frequency bands; the Above 890 proceeding should be viewed in terms of a series of far broader conflicts over deployment of the electromagnetic spectrum. From this perspective it seems evident that intensified use of microwave frequencies by private users was consonant with aggressive expansion by this same class into many segments of the spectrum. Table 1, showing the total number of land and fixed radio transmitters in the industrial radio services, displays the great surge in private spectrum use in this *one* category (of eight now included in the FCC's Private Radio Services) which, though predominantly oriented toward two-way mobile radio, also embraces some fixed microwave. Other private radio services enjoyed similar expansion throughout these years.

The Above 890 docket, I suggest, can therefore best be viewed in terms of a far broader process of private network development, itself crucial to the more general expansion of business throughout the postwar period. By the late 1960s, it was noted by the President's Task Force On Communications Policy (*Final Report* 1968: Chapter Six, 12), private line services accounted for 15 percent of the

TABLE 1
Land/Fixed Stations in the
Industrial Radio Services 1950-1975

Year	Count
1950	2,765
1954	13,515
1958	35,170
1961	76,065
1966	126,558
1970	177,598
1975	321,121

Source: Federal Communications Commission, *Annual Reports* for stated years.

TABLE 2
Growth of Business-Oriented Services Offered By The Bell System
($ millions)

Year	Total Bell System Revenues	Total Toll Revenues	Private Line Toll Service Revenues	WATS Revenues	PLS+WATS as % Toll Revs
1941	1299	417	28	—	6.7
1946	2087	858	41	—	4.8
1951	3728	1370	81	—	5.9
1956	5965	2220	170	—	7.7
1961	8614	3284	332	20	10.7
1966	12,417	5378	535	257	14.7
1971	18,948	8835	933	591	17.2
1976	33,507	16,399	1400	1863	19.9
1980	50,791	26,133	2326	3724	23.2

Source: Federal Communications Commission, *Statistics of Communications Common Carriers* for stated years. Data for 1980 taken from AT&T (1980 *Annual Report:* 30).

Bell System's long-haul revenues and 40 percent of its long-haul circuits, and continued to grow "at a faster rate than Bell's other offerings." Table 2 shows the development of these business-oriented services, again with the intention of underlining the deepening business commitment to a variety of private network offerings. Progressive elaboration of such networks, including those making use of frequencies in the microwave band, would in turn furnish major corporate users with a vital point from which to exert leverage in future policy proceedings.

PRIVATE TERMINAL EQUIPMENT AND THE PUBLIC NETWORK: THE CARTERFONE

The Carterfone proceeding—centering upon conditions governing use of a device permitting interconnection of mobile radio telephone systems with the public switched telephone network—is often cited as the source of competition in today's telecommunications industry. We shall find, however, that the Carterfone decision built on postwar developments and regulatory decisions in the area of private mobile radio and private microwave. In Melody's words (1973: 1263, 1260), "*Carterfone* was not a precedent and policy shattering decision. It was a reinforcement and strengthening of the direction of past decisions;" Carterfone signaled the beginning of policy debate over telecommunications "at a new level and not a resolution of the issue." While chairman of the FCC Richard E. Wiley made a comparable assertion (Hearings Before The Subcommittee On Communications Of The Committee On Interstate And Foreign Commerce, House of Representatives 94th Congress 2d Session, On Competition In The Telecommunications Industry, 28,29,30 September 1976: 58): "in a series of decisions beginning in 1956 and continuing to the present, the FCC (with the consistent support of the Federal Courts) concluded that important consumer rights were being denied under traditional telephone industry practices. . . ."

In 1968 the Commission found that the Carterfone satisfied an unmet communications need; that it did not adversely affect the telephone system; and that the AT&T tariff prohibiting its use was unreasonable and unlawful (13 FCC 2d 420 (1968), recon. denied 14 FCC 2d 571 (1968)). A crucial legal basis for the decision was found in the Hush-A-Phone case (22 FCC 112 (1957)), which concerned a cup-like device produced by a small equipment firm for attachment to the telephone handset to reduce extraneous noise. A U.S. Court of Appeals here reversed an FCC decision that had taken seven years to render (1948–1955), ruling that prohibition of attachment of private, customer-furnished equipment to the public network was unlawful if the interconnection would be "privately beneficial without being publicly detrimental." In the Carterfone case, the Hush-A-Phone precedent was interpreted broadly as a general policy, not limited to the Carterfone device. For AT&T, which resisted the Carterfone decision vigorously and successfully kept it from being implemented for some years, Carterfone heralded a new and unpleasantly liberal shift in FCC policy because, up to this point, the *practical* impact of interconnection requirements had been of minimal importance. As early as 1960-1961, Melody (1973:1265) points out, Bell agreed with the railroads to permit liberalized interconnection privileges for this select group of large users and thus warded off development of loosened interconnection rules as a general matter.

Carterfone, however, was decided in reaction to the dynamic mix of new technology and heavy demand from major telecommunications users. Table 3 shows the development of private land mobile radio in the industrial radio services alone. It was on the basis of this great expansion *only* that Carterfone necessarily acquired substantial interest and importance for a segment of the corporate user community.

TABLE 3
Industrial Radio Services Portable Mobile Units 1950–1974

Year	Units
1950	33,608
1952	81,418
1954	132,425
1956	252,265
1958	381,421
1961	680,880
1964	1,054,584
1966	1,273,673
1968	1,538,525
1970	1,840,649
1972	1,548,228
1974	2,082,471

Source: Federal Communications Commission, *Annual Reports* for stated years.

The Chairman of the Central Committee on Communications Facilities of the American Petroleum Institute, Frank S. Bird, submitted a vitally significant statement delineating the character of this interest ("In the Matter of Use of the Carterfone Device in Message Toll Telephone Service," Docket No. 16942 Before the Federal Communications Commission, 13 April 1967). After outlining his own impressive technical qualifications, Bird, an engineer for the California Company Division of Chevron Oil, detailed the communications facilities utilized by his company—principally in and offshore of Louisiana and Mississippi. Aside from extensive reliance upon common carrier circuits (costs ran at approximately $120,000 a year), the California Division operated its own privately furnished telecommunications facilities. These included VHF (FM) mobile radio systems in six inland fields "plus radio systems which provide coverage of the marshland and offshore area from which 90% of our Division's oil and gas is produced in south Louisiana" (Ibid.: 3). Offshore operations required yet more elaborate instrumentation, including eight multichannel microwave hops and four single channel VHF circuits. In all, Bird claimed, private telecommunications operations embraced about 380 VHF (FM) radio sets, 75 aeronautical radio sets, 7 low frequency aeronautical radio beacons, 16 marine radios, 16 microwave transmitters and 7 radar equipments (Ibid.). These communications facilities were part and parcel of exploration, drilling and pipeline operations; expertise derived from their installation and use permitted the Division to provide "consulting service in the design of radio systems to support specific domestic and foreign drilling and exploration operations in various areas remote from more stabilized production operations" (Ibid.). Again we find the interlock of user and innovator roles and, again, this coalescence formed a basis for exertion of powerful pressure on the evolving nature of U.S. telecommunications policy.

Emergence of such elaborate private customized systems would be especially assisted, Bird believed, by developing easier techniques of interconnection with the public switched network. The California Division was apparently "quite dependent" on outside services in daily operations, "and necessarily must contact outsiders to arrange and coordinate for these services" (Ibid.: 4). The production of crude oil and natural gas was an activity "which by its very nature continues 24 hours a day without holidays and, hopefully. . .without shutdown.

> This makes it necessary that our own specialists, supervisors, and administrators be contacted from persons in the field at times when they are away from their offices and are in their homes. The requirements which cause communication contacts to outsiders or to our own people away from their offices are urgent and the decisions which must be made are important both for financial and safety reasons. It is extremely important that the exchange of intelligence be efficient and accurate. All this creates a need for some type of interfacing between our internal and privately owned communications network and the general exchange network of the common carriers. (Ibid.).

Bird stressed "efficient and accurate" information because, lacking a Carterfone-type device, oil company employees were resorting to copying down information gained from the private system and thence relaying it through separate conversation over the public telephone network. This translation process necessarily involved an intermediary operator, making it both less timely and more prone to error than a direct link between public and private networks. To Bird, thus, "the principal issue of this inquiry is the degree of handicap which is going to be placed upon us in the transmission of intelligence from one communications network, in existence and authorized by law and regulation, to another communications network, similarly existing and authorized by law and regulation" (Ibid: 6).

What, recall, was new and unprecedented about this appeal was its dependence upon already extensive private networks, authorized and expanded between 1949 and the Carterfone proceeding. Having built private systems, petroleum company users found a powerful point from which to argue their case. Indeed, it was the existence of private networks alone which allowed the American Petroleum Institute to assert (Ibid: 4 August 1967: 8) that the telephone companies' opposition to the Carterfone was the stance of "protected monopolies and there is a strong public policy against permitting an extension of the monopolistic powers bestowed upon them beyond their lawfully circumscribed area." The very development of private networks by means of the Special and Safety Radio Services (now Private Radio Services) authorized by the Commission forced the interconnection issue into the regulatory arena—for while interconnection would assist users, its prohibition by the carriers gave the latter de facto control over "areas beyond the telephone" (Ibid.). Therefore, API declared (Ibid.: 10–11),

> the lesson to be learned from this case goes far beyond the particular device in question which happens to be the Carterfone. It is that the Commission should, indeed it must, compel the carriers to remove from their tariffs every semblance of authority to prohibit the use of customer-owned devices in the first instance regardless of their nature. The Commission should give short shrift to any plea by respondents that this might lead to dire results on a record which so clearly shows the abusive and rapacious nature of their conduct heretofore.

This last comment was directed against the claim of the telephone companies, that the Carterfone threatened the technical integrity of the public switched network. Eventually, the FCC found the carriers' claim untenable, and held that a tariff would be unlawful if it assumed a priori that the attachment of customer provided equipment would cause technical or economic harm to the network.

Writing in support of API was the Vice President of Shell Communications, a wholly owned subsidiary of Shell Oil engaged solely in providing private radio communications services on a nonprofit basis to its parent—and one of the largest holders of VHF and UHF authorizations granted by the FCC in the Industrial Radio Services. His firm, said Ray Ransome (Ibid.: 13 April 1967: 1), operated about 2500 mobile units and 340 associated base stations. These were dispersed through the United States and its offshore waters in various configurations. Additionally, Shell held authorizations for some 150 point-to-point microwave stations and 600 other stations in Maritime, Industrial and Aviation Radio Services.

Ransome argued that in some circumstances utilization of a Carterfone-like device "represent(s) the only practical means of meeting the total communications requirement" (Ibid.: 2). This was so because a range of oil industry operations, particularly drilling, production and pipeline operations, "of necessity have moved farther and farther into the remote hinterlands of the nation and even many miles offshore on the high seas." In these areas, public telephone facilities "all too frequently even today are non-existent or so limited as to be virtually useless." Therefore, "primary reliance" in many instances need be placed on oil-company provided communications facilities, especially in the initial stages of drilling, production, pipeline and plant construction. Within 50 or 100 miles of the job site, "reasonably good local and toll telephone service" usually could be had. At these remote locations, therefore, base stations were set up, with sufficient mobile range to cover the desired area, or with a radio link controlling a base station nearer the actual job site, and with access to the public switched telephone network.

For while much of the communication to and from drilling or construction sites might be handled privately, "a substantial portion cannot" (Ibid.: 2–3):

> It simply is not feasible to make each such location self-sufficient either in skilled manpower or in specialized services and equipment. For very practical economic reasons these must be readily available when needed but not left idle for the comparatively long periods when not needed. Thus they must be scheduled closely and freed as soon as possible to move on to the next job. In the case of a drilling project these services and skills are usually required in connection with drilling tools and materials, cementing of casting, directional drilling, electrical logging and testing of the holes at various depths, perforation of the casing in carefully calculated patterns at depths which can only be determined as drilling proceeds, changing and selection of drill bits, and many others. Most of these services are supplied on a time and materials basis by companies organized specifically for the purpose of supporting the petroleum industry's activities.

This whole series of technical decisions, moreover, was intertwined with vital *management* choices, "high level decisions which can only be made on the basis of highly accurate information developed at the well as the projected depth is approached, or interesting formations are encountered while drilling." These, the crucial make-or-break decisions, "are usually made by thoroughly experienced personnel available only in a higher level office location of the company" (Ibid.: 3). Centralized management control therefore urgently demanded transmission of accurate production information without delay between well site and head office and in some cases also to outside services companies as well. Communications, fundamental to corporate decisionmaking, confronted a pressing problem: "how best to transfer such information as this from the privately owned mobile radio system where it originates to the public telephone system which can convey it to its ultimate destination" (Ibid.). Carterfone furnished a far better means of achieving this objective than did other techniques, which included repetition of the message by an operator-intermediary (resulting in frequent garbling and time delays), and loudspeaking telephones provided by the carriers, but which the carriers viewed as subject to tariff infringement when held directly to the private radio system (Ibid.: 4–5). Thus the oil companies were "pleased" when the Carterfone and similar devices were introduced; "It is much more convenient for our operators to use and is capable of even higher quality transmission than the loudspeaking telephone" (Ibid.: 5). The issue of interconnection eventuated in a seemingly irresistible argument for the users, because the telephone company quite clearly might not protest against the manual transfer of information between the two legally licensed systems and was therefore placed in the position of opposing "an efficient, rapid, inherently error-free, and modern method of transferring exactly the same information between the two systems"—and one which ultimately threatened no technical harm to the public switched network. Such an argument, again, would have been infeasible, had the oil companies not invested in massive private telecommunications systems to meet their individual production-related needs.

Energy companies were not the only plaintiffs to use this argument. The National Retail Merchants Association (Ibid.: 4 August 1967: 2) produced its own brief in the case, claiming interest not only in use of a Carterfone device for private land mobile radio, but also "in the future use of the Carterfone as it may be adapted for other types of private communications systems now being used by retail department stores such as in-house private telephone systems." Keeping an eye out on FCC matters relating to member firms' plans for expansion, NRMA "sincerely believes that it would be a serious mistake to view the Carterfone only in its relationship to private mobile communications" (Ibid.). To NRMA, the interconnection issue was far broader in scope—"the particular type of private system utilized with the Carterfone is of no relevance" (Ibid.: 3):

> The principle to reaffirm in this proceeding is the right of a telephone subscriber to use his telephone as he sees fit, granting of course that such use does not technically impair the telephone network (Ibid.).

Therefore, there should be no "artificial limitations" placed upon the use of the Carterfone "or indeed other devices performing similar functions" insofar as the type of private communications system with which they were associated was concerned (Ibid.). As private networks proliferated to suit varied user requirements, rather, the telephone company should be forced to grant interconnection between public and private systems. Underlying the Carterfone case was the principle that any device "that suits a subscriber's convenience" and caused no technical harm "cannot be prohibited by the telephone companies":

> Such a prohibition, if permitted. . .would allow the carriers to wield a gigantic economic club against both subscribers and private manufacturers of miscellaneous convenience devices to effectuate an unlawful expansion of the privileged monopoly area granted to the carriers. Indeed, the burden of showing the necessity of prohibiting the use of such devices must be on the carriers (Ibid.: 4–5).

The conflict between carriers and corporate users now plainly involved critical management decisionmaking powers over production, as choices concerning resource allocation, plant location, market strategy and so forth, were immediately impinged upon by the ability freely to configure telecommunications circuits and facilities. The Carterfone proceeding pointed to a situation that, "as a matter of public policy . . . can no longer be tolerated. To construe the foreign attachment tariff provisions as the carriers would have us . . . . violates the integrity of the user's domain" (Ibid.: 7). This might be so only because the users' domain had become progressively larger during the postwar boom.

Although as has been mentioned, the Carterfone verdict merely raised the issue of interconnection or "foreign attachment provisions" to a higher level of discussion, while AT&T made aggressive use of the court system to alter or delay its impact—successfully enough to defer any easy form of interconnection into the latter half of the 1970s—still, in its wake there arose an entire new industry of "interconnect" suppliers of terminal equipment (Brock, 1981: 242–252). Mushrooming from virtually nothing to close to one billion dollars in sales by 1980, the interconnect companies have employed the most up-to-date computer technology to create ever more sophisticated, specialized, versatile and cost-efficient kinds of terminal equipment. Private branch exchanges—computerized switchboards which today facilitate private control of telephone, data, facsimile and other incoming and outgoing signals—are a market in which AT&T's share is said to have slipped to a mere 54 percent (of a total of $1.94 billion in 1980), even as the absolute size of the market continues to increase ("New Glamour for the Office PBX," *Business Week*, 13 April 1981: 122). Though the degree of competition actually present in terminal equipment markets may easily be overstated by such seeming trends, and the market power retained by AT&T correspondingly undervalued, it is indisputable that the past decade has been marked by steady increase in the range of terminal equipment available and in the number of suppliers of such specialized instrumentation.

This fact leads at once to another—that the hidden agenda of the Carterfone proceeding was dictated by the mounting convergence of telecommunications with data processing. When large users demanded permission to interconnect any kind of customer provided equipment to the public network, so long as it did not threaten the latter with technical harm, they were actually intent on throwing open the doors to telematics. This claim requires intensive discussion in its own right.

THE FIRST COMPUTER INQUIRY

The Carterfone decision initiated resolution of issues far broader than those it explicitly confronted. A hidden agenda of the Carterfone proceeding was formulated, structured and discussed even as that proceeding unfolded, within a far more extensive and wide-ranging inquiry, "In the Matter of Regulatory and Policy Problems Presented by the Interdependence of Computer and Communication Services and Facilities" (FCC Docket No. 16979, 10 November 1966, 28 FCC 2d 291). In this, the First Computer Inquiry, policy implications of emerging user needs for private computer systems harnessed to telecommunications circuits were addressed by regulators for the first time in a sustained and careful fashion.

By the mid-1960s, as third generation computers utilizing integrated circuits began to be installed, the business computer market had become quite substantial. While the sales value of computers installed by 1966 approached $8 billion, the compound growth rate in dollar value of this installed base leapt up 41 percent per year and—reflecting tremendous technical advance—the growth rate of installed capacity shot up 65 percent per year between 1955 and 1965 (Brock, 1975: 24). A survey conducted by the Business Equipment Manufacturers Association, and cited in a Booz, Allen and Hamilton study submitted as evidence into the First Computer Inquiry (FCC, "In the Matter of Regulatory and Policy Problems Presented by the Interdependence of Computer and Communication Services and Facilities," 5 March 1968: 2), showed that the number of non-defense computer systems in use in the U.S. had increased from 91 in 1953, to 427 in 1955, to more than 35,000 in 1966. Distribution of this installed base by user industry may be studied in Table 4; the latter reveals clearly that computers had penetrated virtually the entire industrial and financial base of the United States by the mid-1960s.

Operation and use of this installed base, however, was undergoing profound change. While in 1960, only some 31 computer systems were installed on-line—meaning that use of these computers might occur from remote terminals connected by telecommunications links, either in "batch" mode (implying substantial turnaround time) or in "real time" (meaning that computer processing might take place with such speed as to permit its insertion into ongoing operations—e.g., airline ticketing)—by 1966, the BEMA survey revealed, more than 2300 on-line systems were in use (Ibid.: 16). The number of data terminals installed on-line, moreover, had increased tremendously since 1955, moving from 520 in that year to 45,663 in 1966 (Ibid.: 23). Computer applications, in other words, had come to depend more and more upon communications links, including privately-owned microwave hookups, dedicated leased lines and regular common carrier circuits.

Usage of computerized data processing, and widening efforts to integrate computers into business on a companywide scale, bred further dependence, and it was immediately evident through their filings that major corporations had already grown acutely reliant upon telematics by the mid-1960s. Thus the following commercial users of data systems responded to the Commission's initial invitation to describe current and anticipated uses of computers and communications channels (28 FCC 2d 306):

Commercial Data Systems Users Filing For FCC Docket No. 16979

Aeronautical Radio, Inc.
Aerospace Industries Association of America, Inc.
Aetna Life and Casualty Co.
American Bankers Association
American Business Press, Inc.
American Newspaper Publishers Association
American Petroleum Institute
American Trucking Associations, Inc.
Association of American Railroads
Credit Data Corp.
Eastern Airlines, Inc.
Humble Oil & Refining Co.
Lockheed Aircraft Corp.
McGraw-Hill, Inc.
National Association of Manufacturers
National Committee for Utilities Radio
National Retail Merchants Association
Societe Internationale de Telecommunications Aeronautique
Union Pacific Railroad Co.
United Airlines, Inc.

The Booz, Allen and Hamilton study identified more than 800 individual applications of computers within the industries shown in Table 4 (Ibid.: 25). Approximately 70 percent of computer systems installed by 1966, however, were located in airlines, banking, computer service bureaus, education, the Federal Government and manufacturing. It will be instructive to sketch the varied nature of such applications.

In airlines, computers had materially increased the efficiency of managing and allocating seat reservations and cargo to maximize aircraft utilization during a period of rapidly expanding use (Ibid.: 27–38). Future applications would embrace automation of certain maintenance functions, ground logistics, and so forth; already the number of cities served by on-line reservations systems varied between 4 (Northeast) to 16 (American) to 105 (United). Airlines were among the first to employ data communications on an integrated industry-wide basis.

The National Retail Merchants Association, a trade group representing 15,000 retailers, filed to state that "(t)he interest of the retail industry in the computer is manifold with myriad applications" (Ibid.: 5 March 1968: 1). Retailers like Sears, Ward, Penney and the Singer Company were "making decisions requiring data processing services," chipped in attorney William H. Borghesani later, during oral argument (Ibid.: 3 September 1970: 95). Most retail applications

TABLE 4
Estimated Distribution Of Computers Installations By User Industry (Percentage)

	1953	1959	1966	1979
Agriculture, Forestry, Fisheries	—	—	.2	
Mining	—	.3	2.0	
Construction	—	.4	.4	2.1 (3 categories)
Manufacturing	18.7	42.4	36.8	24.2
Transportation, Communication, Utilities	3.3	8.2	7.9	10.1
Wholesale & Retail Trade	—	1.3	6.9	8.1
Finance, Insurance, Real Estate	1.1	9.9	17.3	18.3
Services (includes Education	23.1	14.6	14.5	24.1
Government	53.8	22.9	13.4	13.1
Nonclassifiable	—	—	.6	—

Note: Percentages for 1953, 1959 and 1966 are for *number* of installations; for 1979 the percentage is for estimated *value* of installed computer base. Definitions of categories may fluctuate across given years.
Federal Government figures do not include all computers employed for military purposes.
Source: 1953 Survey of Automatic Digital Computers—Office of Naval Research; 1959 Census of Installed Computers—General Electric Corp.; 1966 International Data Corporation Census; All in FCC Docket No. 16979, "In The Matter Of Regulatory And Policy Problems Presented By The Interdependence Of Computer And Communication Services And Facilities," Business Equipment Manufacturers Association Comment, 5 March 1968, as supplied by Booz, Allen & Hamilton: 2778–2779. 1979 Standard & Poor's Industry Surveys "Office Equipment Systems and Services," 26 June 1980: 014.

apparently fell under two broad categories: inhouse data processing and electronic data processing services rendered to the retailers by service bureaus.

The Association of American Railroads filed to assert that 235 computers installed in the railroad industry were used primarily for inhouse purposes under immediate railroad control (Ibid.: 5 March 1968: 3). However, the Association continued (Ibid.: 4)—depending in part upon 21 railroad companies' 14,343 route miles of microwave transmission facilities, the railroads' "ultimate goal" was a "total information system" designed to provide "the broadest range of operational informtion to all management levels on a timely basis" (Ibid.: 5). Accounting and record-keeping systems were "continually evolving" in the fields of payroll, stockholder records and dividends, freight car and per diem accounting, inventory control, car repair accounting, and so on. A considerable number of these applications, it was thought, would "naturally gravitate from routine batch processing to on-line real-time operations" (Ibid.: 6).

Conversion of railroad engineer computational work to interactive, on-line data systems, would be helpful to development of grading quantities based on roadbed configurations, grades and contour elevations; to simulation and prediction of performance of different train configurations; to design of bridges and calculation of curve alignments; to scheduling of railroad construction work; and to selection of routes for varied loads on a real-time basis (Ibid.: 8). Computers were being integrated into marketing of freight services, so that selection of industrial sites might draw upon data banks of demographic and economic information maintained about communities served by the railroad and "compar(e) for selection these characteristics against the unique requirements of specific site development projects" (Ibid.: 7). Statistical analysis of current traffic patterns would likewise feed into projections of future trends. The opportunity for extensive market research was widened dramatically by the computerized capacity to rely upon up-to-the-minute data. Switching of trains, prompt location and placement of freight cars, on-line views of an entire freight transport operation including types of cars, ownership, location, load status, destination and various operating conditions—all such data promised "a greater degree of certainty" in scheduling shipments and deliveries. "To the railroad, it means better car distribution and utilization" (Ibid.: 10). In a word, communications were once again a limiting factor in planned corporate expansion.

An analogous pattern was apparent in the comments of the American Petroleum Institute's Central Committee On Communications Facilities (Ibid.: 6 March 1968). The variety of current uses for computers and communications in the petroleum industry was already so great, noted the API, "that this reply will be restricted to illustrative examples" (Ibid.: 2). Moreover, the variety of telematics uses would increase "many fold" over the forthcoming decade: "the contribution of these uses to the efficiency and productivity of the industry will increase at an even more rapid rate, and data communications will become a substantial fraction of the total petroleum industry communications load" (Ibid.: 5–6). Data processing was already indispensable to oil exploration in seismic surveys; computers were alike employed for accounting and finance (payroll, sales records, accounts receivable and payable, stockholder records, special economic evaluation, etc.). Process control and evaluation in refineries and petrochemical plants had become routinely dependent upon computers; so had warehousing and inventory and billing, all of which were further reliant upon data processing to develop both statistical analyses and market forecasts. Petroleum information systems designed to aid management decisionmaking through facile manipulation of massive amounts of, say, exploration data, were again reliant on computers with remote access terminals (Ibid.: 7). The production and transmission of oil and gas utilized data sensors and remote control devices located at well heads, at various storage tanks, and at inlets and outlets to pipe systems and pumps.

> The data signals are transmitted by communications circuits to a central point where a computer processes the data and then transmits, over the communications circuits, control signals to the various control devices and pumps. The

computer is programmed so that the resulting production and distribution are the most efficient that will meet pre-assigned constraints. Moreover, the computer recognizes abnormal conditions such as line breaks, initiates the necessary precautionary action and informs the proper operator concerning the location and nature of the trouble (Ibid.: 7–8).

Long-range planning held that these and other computer-based processes soon would be "interlinked by communications media to provide total corporate information and control systems." (Ibid.: 8).

Still another aspect of the API's concern over the course of computer communications referred to an incipient offshoot of their growing investment in data processing facilities—specialized computer services. Because computers come in discrete sizes "it is highly probable that a computer which has adequate capacity to meet a company's needs for a reasonable time in the future will have capacity in excess of the current needs. It, therefore, becomes economically desirable to make available to others some of the excess capacity" (Ibid.: 8–9). For this reason—that is, to make maximally efficient use of their capital investment—a number of petroleum companies already were providing computer services as a by-product of their involvement, and this too increased their dependence upon communications circuits.

The Aerospace Industries Association of America, with 1967 sales of over $23 billion by member companies such as Boeing, General Dynamics, GE, GM, Grumman, Honeywell, Hughes, IBM, ITT, Lockheed, McDonnell Douglas, RCA, Sperry Rand and Westinghouse, characterized its reliance upon computers as "constantly expanding both in terms of amount and variety of uses" (Ibid.: 6 March 1968: 3). For AIA, further, "(e)fficient use of computers is dependent upon communication facilities" (Ibid.: 3). By virtue of its history, organization and current nature, the aerospace industries were geographically dispersed.

> Typically the administrative offices of an aerospace company, together with some production facilities are located in one section of the United States, other production plants are located in one or more other sections, testing facilities may be in still other sections and subcontractors are widely scattered throughout the country. Communication by wire or radio is essential to the use of computers — frequently situated far from the source of data input and the place where the output will be used—in order to shrink problems of time and distance which affect the efficiency of the aerospace industry. (Ibid.: 3–4).

"There is no business of any significance in the United States," AIA stressed, "which does not use and is not dependent upon" communications links (Ibid.:7).

Following Alfred D. Chandler (1962), Herman (1981: 104) claims postwar years saw a dramatic change in large corporate organizations, which has facilitated centralized evaluation and administrative control over companies having both wider geographic scope and greater vertical integration than ever before.

Functional organization, where major administrative needs (planning, operations, purchasing, finance, marketing) were formalized within corporate structure, was rapidly replaced by divisional organization where, for each product line, a subdivision performed all major administrative functions. Top management, in the newer divisionalized structure, was able to compare and contrast performance between product lines, and might treat each subdivision as a quasi-independent profit center competing with all of the others.

Corporate administration, in turn, rested increasingly upon computer communications. Management information systems were a growing facet of corporate life; built upon data base management they were intended to allow companies "to stay competitive and to improve control over . . .operating environments," as the Electronic Industries Association put it (FCC, "In the Matter of Regulatory and Policy Problems Presented by the Interdependence of Computer and Communication Services and Facilities," 5 March 1968: 36). Management information systems were already in place at Westinghouse, Weyerhauser, Armco Steel, and Dow Chemical—whose Control Data computer in Midland, Michigan, was linked to Dow plants, warehouses and sales offices throughout the U.S. and would, Dow hoped, ultimately "extend throughout the world" (Ibid.: 37–39). Generalizing, the National Association of Manufacturers asserted (Ibid.: 4 March 1968: 1) that both the expanding national economy and this growing complexity of business were "forcing American industry to computerized operations." Business information systems were part and parcel of the evolving trend to greater centralized corporate control; NAM members were being affected by computer communications "at every level of business activity" (Ibid.). NAM advised, therefore, that "the development of computer/communications systems should proceed under policies and procedures most favorable to rapid progress in this important area" (Ibid.: 1–2).

The American Bankers Association, a trade group comprising 13,600 commercial banks, rounded out this series of comments from virtually every major sector of finance, transport and manufacturing. Serving as the nation's principal depository, loan source and payment and transfer mechanism, the ABA's members likewise provided essential financial services incidental to these foremost banking functions. That members might continue to furnish these indispensable services, ABA asserted, and in order to improve and enhance their efficiency, banks had moved rapidly to embrace computers (Ibid.: 4 March 1968: 1). Approximately 1000 commercial banks operated on-premises computers, while 2000 more utilized off-premises computer services furnished both by other, generally larger, banks, and by commercial service bureaus. Demand and savings deposit accounting, loan and trust accounting and internal management and operations were alike being automated. Many banks had begun to furnish additional automated services to other banks and to customers—including payroll accounting, account reconciliation, correspondent bank services, labor distribution, accounts receivable and professional billing (Ibid.: 1–2). It had become feasible to eliminate physical transport of checks, because electronic data transmission by wire or radio permitted a distant user access to a computer as effectively as if the latter were actually in the user's building. Because it had become feasible, computerization was made necessary by the constant press for effective cost control, and this in turn dictated "more and more extensive and more and more effective communications—between bank branches and their head office, between affiliated banks and their holding company, between banks and their correspondents, between banks and the Federal Reserve System and other Government agencies, and between banks and their depositors, borrowers, and other customers" (Ibid.:3). Future planning made adequate communications still more vital. The ABA was convinced that automated banking services, including particularly electronic funds transfers, would "continue to develop rapidly." Further growth depended upon "constant improvements and developments in the related communications facilities and services, at more and more economical rates and charges" (Ibid.: 23). In the adjacent field of insurance, Aetna Life and Casualty concurred with this conclusion (Ibid.: 2 February 1968: 1). "We anticipate a much more integrated use of computing and communications in the next decade . . .Our major data processing efforts will rely heavily on communication facilities."

Two paramount trends, each pointed out repeatedly, although in widely varying industry-specific form, underlay much of the testimony given in the First Computer Inquiry. First, in the Electronic Industries Association's terms (Ibid. 5 March 1968: 25), "it is the computer users who will be developing advanced applications which will be so completely dependent upon the exploitation of the fullest capability of both computers and communications." The corporate user community was integrating telematics into the structure of present and planned business operations at breathtaking speed. Such innovations engendered, of necessity, an evolving alliance between computer equipment and software suppliers and major telematics users, as the two groups cooperated to create ever more encompassing and ever more specialized data processing applications. Second, as was already implicit in this first point, as computers achieved this ubiquitous corporate presence communications was quickly becoming "the limiting factor in the planning of new business systems" (Ibid.: 2 February 1968, Comment of Aetna Life and Casualty: 3). As we shall find below, the nature of the constraints imposed by the communications half of the telematics field was twofold, embracing both technical features of communications systems and basic institutional characteristics of the regulated monopoly service that was U.S. communications.

Before moving on to consider these constraints in greater detail, it may be recognized that the First Computer Inquiry afforded what was in all probability the most general and encompassing high-level discussion and reevaluation of telecommunications policy to have taken place since the AT&T antitrust case of 1949–1956 and, arguably, since the prewar years. In reading the testimony given in the proceeding it becomes quickly apparent that, far from being a parochial discussion of capsule issues, the First Compute Inquiry afforded an opportunity for expression of basic corporate concern over the progressing direction and shape of U.S. telematics. The American Petroleum Institute—always active in this sphere (Ibid.: 6 March 1968: 2)—held that the proceeding had become a matter of pressing necessity:

> The ever growing importance of communications in every phase of our national life continues to raise serious regulatory and policy questions which require timely and informed resolution by the Commission to facilitate the orderly development of all phases of our national life unhampered by lack of adequate and appropriate communications facilities and services made available at reasonable charges and under reasonable terms.

Thus what arose, usually at a governmental level, throughout much of the world only during the late 1970s, and even then by and large only in response to aggressive U.S. involvement in telematics, was articulated and debated by the U.S. private sector fully a decade earlier: How should policy be set to expedite the computerization of society? IBM submitted a list to the Commission (Ibid.: 6 March 1968) of "Data Processing Applications In The Next Decade Showing Estimated Use Of Common Carrier Lines" which broke down economic society sector by sector (retail, wholesale, education, savings institutions, hospitals, government, petroleum and industrial chemical processing, etc.) as well as function by specific function (auditors' reports, general ledger and payroll, educational testing, etc.) as a means of graphically displaying the penetration by computer communications of everyday life.

While the profound implications of telematics for U.S. society were not lost on participants, however, often the long range societal consequences was subordinated to discussion of telecommunications industry tariffs and technical capacities. It was these issues, indeed, that formed the center of the inquiry, because they were quickly becoming coterminous with the far deeper and wider question of computerization per se. Because the monopolistic telecommunications industry—we shall find—hewed to technical standards at variance with those required by computer systems, conflict between that industry and the vast field of users intent upon incorporating telematics into their operations was virtually inevitable; because telecommunications was a *regulated* industry, a large portion of this conflict necessarily occurred in public, or at least in as public a forum as was constituted by the comments and replies heard by the Federal Communications Commission. It is testimony to the importance of the First Computer Inquiry that the issues explicitly raised therein in large part continued to set the policy agenda for telematics into the late 1970s.

The FCC's Notice of Inquiry specifically asked for comment about a wide variety of technical and tariff-related matters (7 FCC 2d 11-18). Among these were: interconnection of customer-provided facilities, owned or leased, with common carrier facilities, including prohibitions against use of foreign attachments; shared use and resale of common carrier equipment and services; the need for new types of common carrier service offerings felt by the computer industry and its customers; and the current inadequacies of carrier transmission facilities in respect of the accuracy and speed required by maximally efficient use of computers.

At the time the First Computer Inquiry began in November, 1966, interconnection and foreign attachment issues were set for extensive discussion in the parallel Carterfone proceeding—as we have seen. Their inclusion into Docket 16979, however, meant that the FCC was explicitly widening the field of legal discussion to encompass several related issues. We have examined user participation in the Carterfone case, and seen how uniformly users held out for a broad interpretation of foreign attachment provisions. Such a liberal rendering would facilitate both attachment of innovative and specialized terminal equipment to common carrier circuits, to meet increasingly diverse and user-specific needs, and interconnection of common carrier circuits with sometimes extensive private network facilities. Users assumed a predictably similar stance in the First Computer Inquiry. Yet because the latter's warrant permitted broader terms of reference, users were able to launch more forceful arguments here than had been feasible in the constrained Carterfone proceeding. In the eyes of the National Retail Merchants Association, for example (FCC, "In the Matter of Regulatory and Policy Problems Presented by the Interdependence of Computer and Communications Services and Facilities," 5 March 1968: 4), Computer I addressed issues "of landmark importance:"

> The resolution of them, in whatever form, will in very large measure determine the use of and the benefits derived from the computer, so dependent is this tool upon the best in communications services.

And, NRMA asserted (Ibid.: 7), the ban on foreign attachment which had served to protect the Bell System was "totally inimicable (sic) to the computer revolution not only as concerns the provision and use of terminal devices, but also of interconnecting devices linking private communications systems with the dial-up network of the carriers." The retailers went on to claim that this tariff prohibition was "nothing more than an illegal tying clause," of the sort declared illegal in a 1936 Supreme Court decision against IBM, which had the effect of "forcing the communicating public to deal with the carriers in areas of services and hardware which are beyond the monopoly franchise of the carriers" (Ibid.). Once more, this assertion might be made only owing to the already massive private corporate investment in both computers and telecommunications facilities. Antitrust implications of the foreign attachment ban, NRMA insinuated, were "enormous." There was simply "no justification— either technical or economic—for the carriers to continue to erect this artificial barrier to a more effective use of their proper services" (Ibid.).

The railroad industry agreed: tariff restrictions on interconnection of common carriers with customer–owned equipment and facilities "should be eliminated" (Ibid.: 5 March 1968, Comment of the Association of American Railroads: 17). As far as foreign attachments were concerned, the Hush-A-Phone court decision spelled out the proper course. "The test should always be whether use of so-called but mislabeled foreign attachments are beneficial to the railroad or transportation customer user and not detrimental to the public interest in the quality of service" (Ibid.). The American Petroleum Institute also concurred; tariffs for common carrier services "should permit, on a compensatory and nondiscriminatory basis, full interconnection of customer-provided computer or message switching facilities" (Ibid.: 6 March 1968: 13). Current constraints on foreign attachments were "outmoded in a data communications environment," chipped in the National Association of Manufacturers, and the appropriate tariff provisions must therefore be revised. Existing limitations on interconnection should also be changed "in view of the whole new context created by developments in the computer/communications systems" (Ibid.: 4 March 1968: 3). Aetna Life and Casualty favored "liberalization" of the foreign attachment rule (Ibid.: 2 February 1968: 2–3); the American Bankers Association was concerned that "undue restrictions on the use of foreign attachments may interfere with the development of the payments mechanism and other financial and incidental services" (Ibid.: 4 March 1968: 22). Should the FCC *not* demand that the tariff be altered, substantive harm might be wrought to the bankers' capacity to serve the public.

Embracing 300 manufacturers and users of electronic equipment, the Electronic Industries Association made perhaps the strongest general argument for revision of foreign attachment tariff bans. Current provisions, EIA declared, were contrary to the public interest because "they deny the full and efficient use of common carrier facilities and services" (Ibid.: 5 March 1968: 121). Future trends and developments would place yet greater strain on businesses seeking to expand, were the tariffs not revised. "(D)ata processing will encompass many new requirements both in the characteristics of data transmission services and terminal equipments," EIA explained (Ibid.). Under existing tariffs, however, "entrepreneurs are deterred from risking capital in the development of equipments which would be considered as foreign attachments" (Ibid.). The carriers, to be sure, had already recognized the need to attach certain types of instrumentation to their transmission system—and thus had leased to customers a few types of devices, most notably, modems, which convert or modulate digital signals into analog signals for transmission of computer data through the telephone network, demodulating the signals back to the digital mode at the circuit's other end. Telephone company provided terminal equipment, however, was simply not sufficiently specialized to meet large user needs. The widely differing character of computer equipment produced by several major manufacturers, coupled with the explosive increase in number of computer applications, mandated the cultivation of a diverse field of terminal devices with a myriad of specific technical features. Each of these small, specialized terminal equipment markets might easily fall below the threshold of commercial interest for the giant AT&T, particularly since voice telephony, its meat-and-potatotes service, still was undergoing tremendous expansion. EIA asserted therefore that it was "not reasonable to expect the carriers to respond to these varied requirements, much less on a timely basis" (Ibid.: 121–122).

These comments were filed by early March of 1968. In June of that year, the Commission ruled that the Carterfone did indeed violate the tariff—but that the tariff itself was illegal. It is thus difficult to conclude that the Commission reached its momentous Carterfone decision oblivious to the arguments and, more specifically, to the groups making these arguments, in the other, larger proceeding.

Interconnection and foreign attachment were but one aspect of the First Computer Inquiry. Corporate users were intensely dissatisfied with the carriers, on a number of counts, and their anxieties were increasing in proportion to the advance of computer system applications. The Business Equipment Manufacturers Association study (by Booz, Allen and Hamilton) identified six interrelated areas in which the services provided by common carriers," do not meet fully the present requirements for data transmission" (Ibid.: 5 March 1968: 167). These were largely a function of the divergent technical features of telephone networks intended for voice traffic, and computer systems configured so as to maximize the economy and efficiency of data transmission. Reflecting the unsolved problem of divergent voice and data requirements were: the range of speeds at which data might travel over common carrier circuits; the classes of modems needed to interface computer equipment with the public switched telephone network; the reliability or error-proneness of the latter; the time required to connect circuits; and rate structures and traffic characteristics of the current telephone network.

Telecommunications circuits are engineered and built to permit traffic to travel at set speeds; this refers not to the pace of the electrical impulse but to the amount of traffic that can move through the circuit in a given period of time. Data users' concern over available line speeds thus reflected the fact that a broad range of computer equipment in commercial use was not matched by an equally diverse field of line speeds from which to choose for transmission of data. The computer user was forced to choose between paying for a considerably higher speed transmission line than was required, or operating his terminal or peripheral devices "well below their most efficient speeds" (Ibid.: 168). A larger number of bandwidths or information channel "sizes" was similarly necessary, were users to attain the flexibility and economy they desired.

For most purposes, major computer users were required to use modems supplied by the carriers when interconnecting with the public switched network. On various counts this was insufficient. Computer systems designed for ever more specialized use did not readily "fit" general purpose modems available from the carriers; and improved economy and performance could result if redundant features presently common to both the terminal and the modem could be eliminated (Ibid.: 171). Were modem and terminal to be combined, moreover, maintenance for both could be provided by one source, which would reduce the time and cost required to identify and correct technical breakdowns (which under current circumstances necessitated repair by representatives of *both* the telephone company and the particular computer or terminal manufacturer). Purchase of combined terminal/modems from equipment manufacturers would facilitate cost–efficient application under predictable circumstances, instead of subjecting the user to unexpected increases in lease charges by the carriers. Finally, as has been mentioned, provision of specialized application terminals with correspondingly customized modems would be economically burdensome for the large carriers owing to comparatively sparse demand. Small data communications manufacturers, on the other hand, might just fit the bill by providing users "with economical, specialized terminal/modem devices" (Ibid.: 172). These manufacturers had already gotten started, significantly, by designing and producing modems on a smaller scale for use on private leased lines and for sale to the carriers themselves.

Greater circuit reliability, also needed for effective data communications, would follow fairly easily if line speeds and specialized modems were improved. Voice traffic, for which the telephone system had been engineered, was not seriously impaired by noise that caused a syllable or a word to be distorted—the listener could usually fill in the gap and supply the meaning himself owing to his grasp of the context in which the sound occurred. For economical and efficient data carriage, however, error rates needed to be far lower, for computers did not have the creative capacity upon which persons speaking with one another routinely relied (Ibid.: 173–174).

Again, rate structures had been devised to fit voice traffic. For computer system users employing long distance service (message toll service), transmission of data was uneconomical. Data traffic tends to travel in "bursts" of high volume; such bursts are of very short duration. Transmission of such data bursts over conventional long distance service circuits was, however, billed at a full three minute minimum charges tailored to characteristic voice telephone use (Ibid.: 175–176).

Finally, the connect times required for efficient carriage of voice and data traffic were also at variance. For an increasing class of on-line computer applications—airline reservations, credit checking, inventory inquiry—the period of time needed to set up and complete conventional telephone circuits was far too long. For persons, such a connect time might take 10 or 15 seconds without annoyance; for computers, whose price/performance was measured in thousands or millions of operations per second, such a wait was a major burden (Ibid.: 176).

For all of these reasons, telecommunications had become a limiting factor in installation and effective utilization of computer systems. It is vital to recognize that the divergence between telecommunications and computer systems reflected a technical disparity, but that this variance only became consequential in light of major corporate users' need for maximally cost-efficient integration of telematics technology. As the National Retail Merchants Association summed up in Oral Argument (Ibid.: 3 September 1970: 95–96):

> Today we are faced with retailers seeking to link via high-speed data transmission circuits several thousand stores, each of which is a remote access point to centralized computers. Such computer communications systems are for inventory and production control, accounts receivable, and payable, credit authorization and other auditable financial transactions where extreme speed and accuracy are absolutely necessary.
>
> We have the computer technology, the personnel, and all else which is needed with one exception, an adequate common carrier data transmission capability. Retailers cannot use some of the present facilities. With these facilities data transmission is too slow to be acceptable. Error rates are too high to be tolerated.
>
> This not only increases costs, but in many cases actually renders the use of sophisticated communications procedures technically impossible.

Other users, each reflecting a discrete set of individual corporate needs, nonetheless agreed with these general sentiments (See Ibid. Comments of: American Bankers Association 4 March 1968: 22; Electronic Industries Association 5 March 1968: 135–136; National Association of Manufacturers 4 March 1968: 31–34; American Petroleum Institute 6 March 1968: 13–16: Association of American Railroads 5 March 1968: 18–19).

It was not, as has been so often claimed, "technology" per se that made tariff and service restructuring essential for large users. Rather, it was the integration of maximally cost efficient data communications within evolving corporate domains. This distinction is vital because, in its absence, analysts may be content to disregard the active, even aggressive, role of users in shaping telecommunications policy, and to substitute an innate, abstract and in actuality nonexistent technological imperative. We have already noted that telematics assisted an ongoing process of corporate restructuring whereby decentralized profit centers could be ever more rigorously monitored by a central management group (Cf. Herman, 1981: 245). Both sides of this complex and neglected process of corporate transformation should eventually obtain further study. At the very least, it seems clear that telematics facilitates decentralized *operations* but, at the same time, may be used *to enhance further centralized administrative control*. A comprehensive survey of distributed data processing recently released by *Fortune* magazine found that while most companies had switched to distributed data processing in order to increase local user responsibilities and to release pressure from the central computer system, they did so precisely in order *to maintain centralized control* over far-flung corporate units (Tim Scannell, "Fortune" Survey Finds No Loss of Corporate Control in Move to DDP," *Computerworld* 23 March 1981: 10–11).

Two points of clarification are in order. First, it was not simply some sort of perverse obstructionism that caused the telecommunications industry—and, in particular, AT&T—to impede the transition to data communications in users' eyes. AT&T was geared to provision of ordinary voice telephone service, which had grown beyond all expected bounds through postwar years, in part owing to the contribution of expanding *business* as well as residential use. Simultaneously, data communications markets, despite their extravagant growth rates, paled by comparison with the continuing surge in voice traffic: such markets were still simply too small in absolute term, to warrant massive changes in tariff structure and technical service offerings. It is therefore not in the least surprising that the Bell System remained aloof from users' demands for technical upgrading and tariff revisions. When one adds to this the fact that users sought concretely to diminish the power of AT&T by their insistence on liberalization of terminal equipment markets and interconnection restrictions, the latter's seeming reluctance to modernize its network to meet corporate user demands becomes all the more comprehensible.

Second, the collision between users intent upon incorporating telematics into developing private networks and the Bell System, which betokened a deepening conundrum over the course of domestic telecommunications policy, was surely debated in private as well as in public. For can it be that investment, financial and other strategic intra- and intercorporate decisions have been *irrelevant* to the evolution of telematics industry structure? Although normally opaque, such decisions should certainly merit investigation in light of their potential impact on, for example, the actual ability of fledgling competitors to AT&T—such as MCI, GTE, and SPC—to marshal the resources necessary to withstand the advance of AT&T itself.

In raising $110 million following careful scrutiny from investment banks, commercial banks and large equipment suppliers, MCI was the largest startup private venture in the history of Wall Street (MCI Communications Corporation and MCI Telecommunications Corporation, In the United States Court of Appeals For The Seventh Circuit, Nos. 80–2171, 80–2288, MCI v. American Telephone and Telegraph Company, Brief of Appellees, 12 March 1981:6). As we shall find upon occasion below, it would seem that for terminal equipment, long-haul transmission and today, possibly, local service, at least a limited measure of competition has been *created and nurtured* by those with means to do so. We must not therefore view such competition as exists as having emerged as a natural condition of rapid technical change and untrammeled economic growth. Competition in telematics comes only as a result of a long series of political and economic decisions, and must be analyzed in these terms.

Let us return to the First Computer Inquiry keeping these points in mind: The largest corporate users employed the proceeding to develop an agenda of key issues demanding evaluation and, they preferred, redress. Elimination of foreign attachment and interconnection bans would result in their obtaining greater capability for adding specialized terminals to public and private communications circuits. Further, it would expand user control over configuration and utilization of a mix of interconnected public and private circuits, thereby furthering broader corporate strategies. Technical upgrading of the public switched network and tariff revisions would aid data communications in like measure as the latter relied upon the public network.

Beyond these already vital issues, users demanded that the Commission compel the carriers to develop tariffs for sharing and resale of communications links. Sharing of circuits would permit them both to design and implement customized networks to serve specific industry groups (the airlines had long done so, and had been able to justify their need on the basis of air traffic safety and efficiency), and to take advantage of significant economies of scale that may accrue. Resale of extra capacity by users would likewise permit them to achieve more efficient exploitation of their extensive private facilities (including leased lines); moreover, resale was critical to the expanding business of timesharing, whereby users and service firms sold remote access to computer processing power. Resale and sharing both became especially desirable in light of an established rate differential between cheap, bulk private line service—Telpak—and far dearer circuit rates found by smaller users. Resale of Telpak by those who could afford it would permit large users with excess capacity to undercut the rates charged by AT&T for service to less favored customers.

Users such as Aetna Life and Casualty, the Association of American Railroads and the National Association of Manufacturers fought for resale and sharing at the FCC (FCC, "In The Matter Of Regulatory and Policy Problems Presented by the Interdependence of Computer and Communication Services and Facilities," 2 February 1968: 2; 5 March 1968: 18; 4 March 1968: 3, 32). The Electronic Industries Association declared that prohibitions against sharing should be relaxed (Ibid.: 5 March 1968: 132) "to permit the suppliers of communications dependent computer services to develop their businesses or services with maximum freedom." Users too were beginning to spin off specialized computer services, as in the petroleum industry where a number of companies had become engaged in computer services as a byproduct of their earlier incorporation of computers. "It may be argued," conceded the American Petroleum Institute (Ibid.: 6 March 1968: 14), "that the resale of communications facilities and services . . .places the original customer in the position of competing with the communications common carrier through the use of the carrier's own resources and that this is somehow abhorrent." Both of these propositions, API asserted, were "irrelevant" and of "dubious validity" (Ibid.). Present restrictions on resale were "highly unrealistic and strongly characterized by arbitrary, capricious and inconsistent interpretation and enforcement" (Ibid.).

> When the use of communications facilities are (sic) incidental—albeit essential—to the rendition of the original customer's normal business service, no restriction should be placed on such use of these facilities. The case in point of remote access computers clearly demonstrates the need for the removal of these restrictions. Suppose the computer service company has a large central computer in city A and a small computer in city B connected to

the large computer by circuits leased on a full-time basis from a communications common carrier. Suppose a customer of the computer company in city B requires a service which involves both computers. The communications circuit, although essential to the operation, should certainly be available for this purpose. This operation, clearly in the public interest, would be completely inhibited unless this were the case. Any alternative that might be proposed will be either impractical or highly wasteful of communications facilities (Ibid.: 14–15).

Timesharing—multiple access to computers through communications circuits to remote locations—was already undergoing great expansion and, various parties protested, was clearly going to experience rapid growth into the future. The Association of Data Processing Service Organizations (ADAPSO) therefore strongly encouraged "elimination of arbitrary restrictions on the expansion of services" (Ibid.: 5 January 1968: 13). (Also see Ibid.: Comments of the Business Equipment Manufacturers Association 5 March 1968: 103; Electronic Industries Association 5 March 1968: 29).

Users did not win relaxation of provisions governing resale and shared use of communications links as a direct result of the First Computer Inquiry. Indeed, they did not gain capacity to engage in unlimited resale of sharing until virtually the time of this writing. What they doubtless *did* achieve was placement of resale and shared use issues on the unfolding agenda faced by regulators and policymakers. Users effectively served notice on the FCC that these matters should be put on the docket and this by itself must be classed as a victory for the user community.

The most crucial issue addressed in the First Computer Inquiry, however, and the one that we have not yet discussed, was a matter on which the entire user and equipment supplier alliance took a steel-hard stand. In a nutshell, this was whether data processing and computer services which involved use of communications facilities should be deemed subject to regulation as common carriers pursuant to provisions of the 1934 Communications Act. Related to this issue was the question of whether the Commission's existing regulatory controls were sufficient to insure fair and effective competition between carriers and other entities engaged in sale of computer services dependent upon communications.

The status of government regulation was the key issue in the proceeding, the question around which all others turned. Possible extension of regulation would impinge directly and dramatically upon private ownership and control of telematics. Convergence of data processing with telecommunications had already made it exceedingly unclear where one began and the other left off. As the American Petroleum Institute aptly stated (Ibid.: 6 March 1968: 3): "What length must a cable be before it ceases to be part of the computer and becomes a communications circuit?" The computer industry—data processing service firms, suppliers and major corporate users—was *unregulated;* telecommunications, on the other hand, was subject to regulation by the Federal Communications Commission. Virtually the whole field of user and supplier groups were relentlessly opposed to imposition of regulatory controls over data processing, and were also deeply concerned about the terms on which AT&T might be permitted to furnish data procesing services. Speaking for the banking sector, the American Bankers Association argued in terms identical to those chosen by other users that use of computers by commercial banks "is not, and should under no circumstances be deemed subject to regulation . . .whether or not involving the use of communication facilities and services" (Ibid. 4 March 1968: 19; see also Ibid. NAM 4 March 1968: 2–3; EIA 5 March 1968: 26, 100; NRMA 5 March 1968: 5–7; AARR 5 March 1968: 15; AIA 6 March 1968: 4–5). The American Petroleum Institute felt that the question of regulation touched directly on private user control over the communications resource and placed it in sequence with past decisions (Ibid.: 6 March 1968: 10):

> The Commission in the past has established a practice of non-regulation of private communications facilities (except for licensing regulation where radio is involved) when such facilities are provided and operated by the user incidental to carrying out the business of the user. This has applied even though the private communications facilities may have been connected to the communications devices and lines furnished under regulated tariffs by communications common carriers. Any change in this past philosophy of non-regulation of private communications facilities, *and computers*, where provided by the user may seriously restrict the growth of computer usage in the petroleum industry.

User hostility to extension of regulation therefore grew directly from a concern to extend the predictability and flexibility conferred by private ownership and control into the burgeoning telematics field. Data processing had already impinged on applications of such importance to the very largest users that the latter, in preference to timesharing, were reported to "feel it necessary to have the computer under their own control" (Business Equipment Manufacturers Association Comment Ibid.: 5 March 1968: 101). To the extent that it raised *any* uncertainty over the terms that would guide use of data processing, the possibility of increased government regulation struck at extension of sovereign private control. In like measure as telematics was incorporated into a company's operations regulation posed the threat of FCC intervention into substantive processes of private corporate strategy. The Aerospace Industries Association spoke plainly for all major users (Ibid.: 5 March 1968: 7–8) when it argued that the 1934 Communications Act conferred no mandate whatsoever upon the Commission to expand its jurisdiction into the data processing field. The Communictions Act was just that—a *communications* act—and "does not confer any authority otherwise to regulate the business" of any industry (Ibid.: 7). Even where computers were utilized in conjunction with radio transmission or with carrier facilities "they are no more subject to regulation under the Act than the voices of telephone users" (Ibid.: 8).

Users similarly hoped to prevent regulated common carriers from furnishing data processing services under tariff. Extenstion of regulation to the whole of the data processing side of the telematics industry would otherwise be an omnipresent threat. Participation by AT&T in data processing, moreover, boded intrusion of a staggering market force into the heartland of the users' private domain, owing to the telephone giant's command of a vast communications network infrastructure. Were AT&T to be allowed to furnish data processing services, in short, it might either do too little—or too much. Thus the carriers' entry into data processing—that is, into provision of integrated telematics services—presented what the API termed (Ibid.: 6 March 1968: 10) "a very thorny problem" How, API asked, "are competitors of a communications common carrier to be protected from unfair practices by the carrier if they both offer computer services and the competitor is dependent on the franchised communications common carrier for the communications facilities and services which are essential for the conduct of his business?" (Ibid.). How were carriers to be prevented from subsidizing their computer businesses by excessive communications charges? How, API asked (Ibid.), "is the competitor to be assured of receiving needed circuits as promptly and with comparable quality as does the computer division of the carrier.?"

If business users tried to forestall regulation and still prevent AT&T from occupying the telematics field on its own terms, they had to do so while also supplying the Bell System with incentive to upgrade the entire national telecommunications network—of which it forms the core. That is, while users wished to limit AT&T's market power, at the same time they insisted that it modernize to accomodate the technical and cost features of integrated data traffic. How could these diverse—even conflicting—goals best be achieved?

The solution worked out to expedite matters was ingenious, important, and patently temporary. FCC forbearance from regulation of data processing might be accomplished through recognition of a basic distinction between data processing and communications. Although such a distinction was already technically obsolete, it was essential to continuation of data processing as an unregulated private activity. The National Association of Manufacturers gave a cue to the Commission in asserting (Ibid.: 4 March 1968: 3): "Data processing and data communications are separate and distinct activities and are identifiable as such." NRMA declared (Ibid.: 5 March 1968: 5–6), "data processing effectuated with the computer is a function separate and apart from common carrier communications services. As such, the retail industry regards the use of communications lines as incidents of data processing service designed to promote the effective distribution of processed data." In its final decision, the FCC accepted such arguments in toto, and disclaimed any intent to regulate data processing services at the present time: "we see no need to assert regulatory authority over data processing services whether or not such services employ communications facilities in order to link the terminals of the subscribers to centralized computers" (28 FCC 2d 298). Computers might continue to be integrated into production and corporate administration as privately owned plant. With this basic decision made, the evolving balance between private user control, and quasi-public control over the domestic telecommunications system by regulated carriers, was tipped dramatically toward the former.

As a result of this critical verdict the FCC was forced to accept a complicated set of distinctions with which to distinguish data processing from communications, thereby staking out the limits of its own regulatory power. Four categories were delineated. The Commission claimed regulatory authority over two—hybrid communications and communications—which were classed as such if the "primary use" of the service being offered was communications. That is, a degree of reliance on computers would not alter the fundamental identity of a service as communications. The problem was, of course, precisely a question of degree. At which point did an integrated computer communications service shade over from "communications" to "data processing"? Predictably, such a clumsy distinction proved eventually to be untenable. However, in large users' eyes, its unwieldiness was far less vital than the benefits which followed from its application.

In addition to erecting high barriers to extension of regulation and thereby in effect tacitly sustaining a massive expansion of private power, the distinction was employed to permit common carriers to organize *un*regulated separate subsidiaries to provide data processing services. Incentive to modernize and upgrade facilities for data carriage was thereby conferred upon the carriers, who might by this means themselves share in the benefits of modernization. This course had been advanced by the American Petroleum Institute (FCC, "In The Matter of Regulatory And Policy Problems Presented by the Interdependence of Computer and Communication Services and Facilities," 6 March 1968: 11), contingent upon "the most careful attention" from regulators as to relations between regulated carriers and their unregulated subsidiaries. The American Bankers Association and the National Association of Manufacturers (Ibid.: 4 March 1968: 21; 4 March 1968: 3–4) likewise endorsed the need for close regulatory scrutiny to ensure that subsidiaries utilized communications facilities and services under exactly the same tariffs as the general public of large corporate users. The Commission acquiesced, and ordained that separate subsidiaries would have to keep separate books and obtain communications from parent companies "pursuant to the same tariff terms, conditions, and practices as are applicable to any other customer of the carrier" (28 FCC 2d 303).

The distinction between data processing and communications, however, ingeniously permitted only the independent telephone companies to become engaged in the former through subsidiaries. Under the terms of a 1956 Consent Decree obtained by the U.S. Department of Justice, AT&T was prohibited from engaging in business that was not incidental to its regulated common carrier services (U.S. vs. Western Electric Co., Inc., and American Telephone and Telegraph Company 13 RR 2143; 1956 Trade Case 71,134 filed 24 January 1956). FCC acceptance of a firm division between data processing and communications, however it set reality ajar, effectively barred the Bell System from provision of those services which were classified as data processing by the Commission. This by no means constituted absolute foreclosure. On the contrary, the very opacity that inevitably surrounded the distinction implied that AT&T had been given room to test its nature and limits by pragmatic means. From the viewpoint of large corporate users, this was all well and good: while unregulated data processing subsidiaries of independent carriers might form a leading edge in front of the Bell System, instilling in the latter an incentive to upgrade its plant so as eventually to meet them head-on, the very murkiness of the administrative fiction established by the FCC in separating data processing from telecommunications gave AT&T additional flexibility in developing data traffic capacity. Users *needed* the Bell System—they still do—because it forms the core of the national telecommunications infrastructure. The debate has been only over the conditions of use to be placed upon the Bell System network, and over the relative power to set these terms held by large corporate users and by AT&T itself.

The verdict returned by the FCC in its First Computer Inquiry followed an earlier trajectory. Centrally administered, geographically dispersed corporate users would be granted more power to deploy the communications infrastructure as they chose; if this expanded freedom came at the expense of AT&T, then so be it. Telecommunications policy, the President's Task Force had prescribed in its Final Report (7 December 1968: Chapter One, 6), "should seek to maintain and develop an environment always sensitive to consumer needs." In succeeding decisions, to which we now turn, we shall find evidence of further efforts by corporate users to expand and upgrade available technical capacities through a policy of enlarged competition in domestic telecommunications.

THE SPECIALIZED COMMON CARRIER PROCEEDING

Looking back on the First Computer Inquiry from the vantage point of 1977, the Federal Communications Commission found that responses to that inquiry had made four points "abundantly clear" (Statement of Hon. Richard E. Wiley, Chairman, Federal Communications Commission, Before The Subcommittee On Communications Of The Committee On Commerce, Science, And Transportation, U.S. Senate 95th Congress 1st Session, 21, 22 March 1977, Part 1, Serial No. 95-42, *Hearings On Domestic Telecommunications Common Carrier Policies*. Washington: GPO, 1977: 98):

> (1) communications facilities and services related to computer/communication services uses were in their infancy; (2) timely development of communications services adapted to these special needs was critical to U.S. economic

growth and needs; (3) a number of users were dissatisfied with many aspects of the telephone communications capabilities, and these users had special communications requirements; and (4) special private line caapbilities (sic) were needed for many data communications applications, particularly high speed data transmissions.

Users indeed had been vociferous in their demands for better, cheaper communications service from the carriers, while Computer 1 had for the first time afforded a general discussion of the scope of the problems involved in satisfying these demands. Owing to the rapidly escalating dependence upon communications by U.S. industry and finance, a single company—AT&T—was in practice materially affecting (some said dictating) the pace and direction of corporate strategy in areas far removed from its traditional communications offerings.

"Alternatives must be found," asserted William H. Borghesani on behalf of the National Retail Merchants Association during the First Computer Inquiry (FCC Docket No. 16979, "In The Matter of Regulatory and Policy Problems Presented by the Interdependence of Computer and Communications Services and Facilities," 3 September 1970, Oral Argument: 96). "One which retailers view with much hope in the private line area is the development of such specialized common carrier systems as had been proposed by MCI" (Ibid.).

After nearly six years of proceedings, MCI had, in August of 1969, received a go-ahead from the FCC to build a microwave system between St. Louis and Chicago to offer voice, data transmission, facsimile and other private line services, in competitition with AT&T (Brock, 1981: 211). The established carriers had bitterly opposed the application, effectively causing their fledgling competitor (and its backers) to expend some $10 million in regulatory and legal costs associated with obtaining permission to construct what was at first a $2 million communications network (Ibid.: 213). According to Brock (Ibid.: 212), a factor influencing the Commission's decision was "MCI's proposed flexibility in providing services tailored to individual customer requirements compared with AT&T's tariff rigidity." MCI could package its service more flexibly than the established carriers to meet specialized user needs *because* it was a small, new company, targeting a specific market and not yet committed to heavy capital expenditures in support of a voice network. As Borghesani put it during the First Computer Inquiry (FCC Docket 16979, "In the Matter of Regulatory and Policy Problems Presented by the Interdependence of Computer and Communications Services and Facilities," 3 September 1970, Oral Argument: 96), MCI "was permitted to enter the field between Chicago and St. Louis for the purpose of providing the public a wider range of voice and data transmission services." Speaking of users in general, Borghesani hoped that existing carriers "like AT&T" would be able to "develop drastically improved data transmission capabilities" (Ibid.: 97). To spur them to do so, and thus to service adequately "vastly compounding requirements for sophisticated data transmission capability," NRMA believed that' (n)ew carrier-to-carrier competition is needed in the data field" (Ibid.). Telpak had, as we have seen, siphoned off some of the business that might otherwise have gone into construction of private microwave systems; private line services were booming. "We see several advantages," claimed the Final Report of the President's Task Force On Communications Policy (1968: Chapter Six, 12), "to making available more potential business opportunities in these markets."

Following the 1969 decision authorizing MCI to construct its link between St. Louis and Chicago, a number of similar applications were filed before the FCC (Brock, 1981: 214). The commission took the occasion to institute a broad policy inquiry regarding the desirability of specialized common carriers, and invited comments from interested groups. Responding predictably, the carriers opposed the new competition on technical and legal grounds. On the other side, in Brock's phrasing (Ibid.), "(t)he potential new carriers, their equipment suppliers, and their potential customers all supported new entry." Of this trio, users have once more been typically undervalued.

It is of quite fundamental importance that FCC Docket No. 18920—The Specialized Common Carrier Proceeding—not be viewed as simply extending newly "procompetitive" policies on some abstract philosophical basis. The proceeding was not about competition, but rather, about its intended *results*: better, cheaper services for heavy corporate users with specialized communications needs. Representing a diverse group of major companies banded together explicitly to achieve telecommunications policy goals, and including Bethlehem Steel, Union Carbide, Dupont, Amax, Chrysler, Weyerhauser and Monsanto, Jeremiah Courtney made this point abundantly clear (FCC Docket No. 18920, "In The Matter of Establishment of Policies and Procedures for Consideration of Applications to Provide Specialized Common Carrier Services in the Domestic Public Point-to-Point Microwave Service," Oral Argument, 22 January 1971: 2967–2968): "We are not shooting for competition for the sake of competition." Rather, there were "values which we see will spring from the kind of competition the Commission has within its power to authorize." What values? "First, and probably most important," answered Courtney (Ibid.: 2968), "the new carriers will offer a service that is not now available, namely, a digital network. This type of dedicated facility will be more responsive to users' requirements because its design will provide higher reliability for data transmission." Both the divergent technical requirements needed for data as opposed to voice traffic and the dismally dissatisfying response thus far received from the established carriers, forced users to demand a means by which to enlarge data transmission facilities.

It is worth pausing to note that this issue of data transmission, which returned directly to the configuration of the domestic communications infrastructure, was not unique to the United States. "I understand that the European countries are about equally divided as to whether they will try to tack onto the existing system or build an entirely new system," said Courtney (Ibid.: 2969): "So other countries have problems too." It was not "competition" that underlay the Specialized Common Carrier decision; it was an urgent need felt by user companies globally for telecommunications facilities capable of supporting their enlarged data transmission requirements.

Similarly, the virtues of competition as an economic doctrine were beside the point. "Quite frankly," conceded the Utilities Telecommunications Council—a group of public and private electric, gas, water and steam utilities (Ibid.: 2964)—"the user could care less what happens to some of the carriers that go in." It was advanced data communications *services* at reasonable prices that were vital—not who provided them. New carriers were called for merely to break a logjam caused by AT&T's technical inadequacies and by the telephone company's reluctance to meet user demands for technical efficiency and economic pricing in data transmission. As Courtney asserted (Ibid.: 2970): "all the activities which the specialized carriers may be permitted to perform should be open to the Bell System, so that there will be no loss of Bell System competition in any service area . . .We want to see them in this competition in every aspect."

"(E)ncouraging the provision of adequate common carrier services for data transmission must be a primary goal for the Commission in this decade," stated Herbert E. Marks for the Computer Time-Sharing Services Section of ADAPSO (Ibid. 29 September 1970: 282). Although participants may have been publicly reticent concerning the concrete implications of such a goal, it was immediately evident that it encompassed decidedly more than what was then being proposed by any particular new carrier. "It is important that this field not be limited to long haul operations," asserted Mobil Corp. (Ibid. 23 September 1970, Letter from F.M. Pelka: 157): "A requirement is a new type local distributive arrangement in keeping with the long-haul techniques which the specialized carriers are proposing." Later, the National Association of Manufacturers wrote (Ibid. 16 August 1971: 3926) to advise the Commission that NAM "has a great interest in the issue of local distribution of the proposed services as well as in the allocation of frequencies for intra-city use," and suggested that William B. Pomeroy of General Electric be considered "the manufacturer-user group representative" for service on an advisory committee on these issues proposed by the FCC.

Meantime, the specialized carrier concept would be "an excellent vehicle for tapping the vast market potential of unique telecommunications applications which existing carriers have not exploited"—felt Mobil (Ibid. 23 September 1970: 156–157). Established carriers had deterred Mobil from realizing its objectives in closed circuit television and video phone service, and in high speed data transmission, by varied means, ranging from "uneconomic pricing" to an "inability to offer the desired service" (Ibid.). "(I)f the delaying tactics of the existing carriers are allowed to postpone this needed dimension in telecommunications services many users will be denied the advantages that could be provided by existing technology."

Also participating in the proceeding was the American Petroleum Institute. "As the Commission is very much aware," API pointed out (Ibid. 1 October

1970: 346), "the Central Committee (On Communication Facilities—DS) has participated in numerous proceedings involving the interest of the petroleum radio users." There was, API discerned, "an increasing need for more modern communications systems, designed to meet the technology of computers and at the same time to be fully compatible with common carrier facilities and private system facilities." That need, already "particularly acute" in the petroleum industry, would be well served by development of specialized communications systems such as those being proposed by the prospective new carriers. Authorization of these proposals, further, "would help foster a competitive environment within which users would have a wide range of choices as to the means of satisfying their diverse communications needs" (Ibid.: 347). This plea for open entry, however, was strictly bounded by a limiting consideration of especial importance to API, and other very large users (Ibid.: 347–348):

> we must emphasize that the entry of the specialized carriers into the communications market must be accommodated within the limitations of radio frequency availability without disturbing the present allocations for private communications systems in the operational fixed band. If users are to continue to have a wide range of choices as to the means of satisfying their communications needs, there must be a viable private systems alternative. If, for example, the specialized carriers upon entering into the field are permitted to monopolize the microwave portion of the spectrum, users would be denied as a practical matter the availability of private microwave systems.

Other user representatives agreed. "(I)n allocating frequencies for radiated systems," NAM declared, the FCC should "recognize the needs of manufacturers and insure that manufacturers as well as common carriers have frequencies for their own use" (Ibid. 16 August 1971: 3927). Computer communications systems were becoming "an integral part of innumerable manufacturing operations." Therefore, manufacturers were entitled to "a free choice of obtaining such services and facilities either through the common carriers, or through utilizing their own equipment, as well as a combination of both" (Ibid.).

The Amerian Bankers Association likewise felt that it was not entry of new carriers per se, but the resultant expected provision of specialized services, that made FCC authorization desirable. As data processing and data communications became more critical to banking, the need for services "specially designed and tailored to meet the needs of the banking industry and its customers" correspondingly increased. From experience, though, the ABA was convinced "that fully adequate communications service, particularly the high quality transmission facilities needed for high speed data transmission at rates about 3,600 bits per second, is frequently not now available" (Ibid. 12 October 1970: 1525). Moreover, further developments in data processing would all too quickly enlarge "very greatly" the need for such increasingly sophisticated and specialized communications (Ibid.). The bankers on this basis asserted that "the most rapid and effective development of the specialized services required for high speed data transmission would be achieved by the encouragement of competition in this limited and specialized area" (Ibid.).

The manufacturers represented at the proceeding by John Sodolski for the Electronic Industries Association were convinced that their opportunities for profit in such a competitive environment "would assuredly lead to innovation and development in areas which have generally not been addressed by domestic communications equipment manufacturers" (Ibid. 22 January 1971, Oral Argument: 2905). Thus the alliance that had been forged between electronics and computer equipment companies and major users grew stronger.

It was the National Retail Merchants Association however, which perhaps presented the most urgent case for authorization of specialized carriers. "As the Commission is well aware," it reminded, the NRMA "is one of the principal industry associations active in common carrier proceedings" (Ibid. 1 October 1970: 372). The "growing dimension of the need of retailers" for technically improved and economically attractive voice and data communications capacity had been "abundantly documented" (Ibid.: 373–374). NRMA found the current proceeding of "landmark importance," and hoped to see the specialized carrier concept implemented "as soon as possible in order to bring a meaningful choice among carriers to retailers who need specialized common carrier service" (Ibid.: 374). And NRMA pointedly remarked that "(t)his need cannot wait" (Ibid.). Were the Commission to approach specialized carrier applications on a case-by-case basis, allowing comparative hearings as to their utility, "many years of delay would ensue" during which the need of retailers for improved communications "would not be met" (Ibid.: 374–375). Clearly, NRMA cautioned, "this is not in the public interest" (Ibid.: 375). The FCC should move with dispatch to finalize authorization of a new *class* of specialized carriers.

During Oral Argument, the basis of NRMA's position became clearer. Data transmission needs in retailing were growing at between 25 and 50 percent a year (Ibid. 22 January 1971, Oral Argument: 2926). Integration of computer facilities created a pressing need for price levels and high speed, error-free data transmission capabilities of a sort unmet by conventional carriers. "The Nation's retailers are fundamentally dissatisfied with the pricing and technical quality of data transmission offerings of existing carriers," complained NRMA (Ibid.). Private line reliability, ordering lead time and data circuit quality were given low marks by the very largest retailers in the group (Ibid.: 2928).

This dissatisfaction with the conventional carriers, NRMA continued, "is what brought the NRMA to the communications scene in Washington in the first place," fifteen years earlier for the Above 890 rulemaking. In the wake of that decision, retailers had been "naively . . .delighted" with the new rates offered in AT&T's Telpak tariff schedule, and therefore had not built their own private systems (Ibid.: 2829). Having been "ensnared" by Telpak in 1961, and having structured their private line budgets around it, retailers had been burned badly when, in

1964, some of the Telpak tariffs were canceled by the FCC for being discriminatory and unjustified (37 FCC 1111, See Brock, 1981: 208). Desiring more attractive technical and cost alternatives, retailers had participated in the Carterfone proceeding; again they had been opposed by AT&T; once more the competitive outcome of that decision had been "most salutary for retailers" (Ibid.: 2829). "The competitive spur as exemplified by private microwave and by the Carterfone proceedings," NRMA summarized, "is the only successful way we have found to give retailers much needed alternatives in the common carrier area. We do not see the specialized common carrier proceeding any differently" (Ibid.). Competition was desired by users not for its own sake, but because it seemed the only way to force upgrading of the nation's telecommunications facilities upon established carriers. "What assurance do we now have," NRMA queried, "that AT&T will go forward and meet this now belatedly acknowledged demand for a better data communicating capacity if it is not spurred to react to the competitive necessity of the market entry of the new specialized carriers"? (Ibid.: 2930).

It is vital that we be clear in linking NRMA's complaint against AT&T to retailers' evolving corporate strategy and planning. The projected growth of new data communications applications in retail environments was related specifically to the advent of point-of-sale devices. The latter's dependence upon cost-efficient and technically advanced high speed data communications placed a material constraint upon their actually becoming "the biggest development in retailing today" (Ibid.: 2927). Point-of-sale terminals would, however, allow better oversight of distribution of goods from regional warehouses to stores, with corresponding increases in inventory control. Point-of-sale services would also reduce store operating costs and provide key sales data to retail chains and manufacturers. They would, moreover, facilitate development of on-line sales and credit systems. The feasibility of such systems, which it was hoped would significantly reduce fraud losses, depended on availability of both low cost terminals and error-free data transmission (Ibid.: 2928).

"We are grateful for your private microwave decision," allowed NRMA to the FCC, "(b)ut we would prefer not to be in the communications business ourselves" (Ibid.: 2930). Telpak had afforded what turned out to be only temporary aid; "we have had two rate increases and threatened termination" of Telpak and, in any case, Telpak was not technically up to the rigorous requirements of retailers. As Telpak was no longer a viable alternative, retailers wished to turn to specialized carriers. Their only other alternative, NRMA projected, was "massive involvement in cost shared private microwave" (Ibid.). Climaxing its case, NRMA finished (Ibid.: 2930–2931):

> Retailers can not wait. We need a creation of a new common carrier industry of a specialized nature to help serve our needs. We need them now, we can't wait another fifteen years.

In May of 1971, the Commission announced a new general policy "in favor of the entry of new carriers in the specialized communications field," justifying its action in terms of "the public interest, convenience, and necessity" (29 FCC 2d 870, 920). Looking back on this decision in 1977, then FCC chairman Richard Wiley took a stance which seemed to have been borrowed, lock, stock and barrel, from large corporate users (Statement Before The Subcommittee On Communications, U.S. Senate 95th Congress 1st Session: 116): "Permitting the entry of specialized common carriers would provide data users with the flexibility and wider range of choices they required. Moreover, competition in the private line market was expected to stimulate technical innovation, the introduction of new techniques, and production of those types of communications services which would attract and hold customers."

As always, it must be cautioned that the Commission had not resolved the matter—AT&T withheld interconnection from the new carriers, thereby damning them to operate in artificially restricted markets and contributing to the conditions which engendered a new antitrust case in 1974 (See Brock, 1981: 210–233)—but its action raised discussion of data communications services and policy to a new level of intensity. Users had forced AT&T back another step and, in the process, had signaled that henceforth they would be reckoned a major force in the policymaking arena.

SATELLITES AND VALUE-ADDED NETWORKS

Two other FCC decisions extended the policy of multiple entry of specialized common carriers and purposefully expanded the gray area of hybrid services drawing upon both telecommunications and data processing. We may briefly assess them here.

The Domestic Satellite decision of 1972 (35 FCC 2d 844) borrowed directly from the rationale offered by the Commission for its Specialized Common Carrier verdict of the preceding year. "Competitive sources of supply for specialized services, both among satellite licensees and between satellite and terrestrial systems," the FCC later noted, "would encourage service and technical innovation and provide an impetus for efforts to minimize costs and charges to the public" ("An Overview Of The Domestic Telecommunications Industry And The Commission's Policies" attached to Statement Of Hon. Richard E. Wiley, Chairman, Federal Communications Commission, Before The Subcommittee On Communications, U.S. Senate 95th Congress 1st Session, 21, 22 March 1977, *Hearings On Domestic Telecommunications Common Carrier Policies:* 117). Restrictions were placed on existing carriers "to prevent them from using their monopoly services to subsidize operations in domestic satellite service" (Brock, 1981: 259–260). The United States signatory to the International Telecommunications

Satellite Consortium, Comsat, was told to form a separate subsidiary if it wished to engage in provision of competitive domestic services. AT&T was allowed to employ satellites for its monopoly long-distance domestic services but would have to hold off for three years after it instituted satellite service before providing satellite-based private line offerings.

The Domestic Satellite decision was, in Brock's words (1981: 260), "a very significant regulatory move toward competition in telecommunications"—not merely because it further opened business communications services to entry by new carriers but also because it effectively promoted development of "an important new technology outside AT&T's control." Again, however, competition offers insufficient explanation of the Commission's unprecedented series of actions. Who benefited from satellite communications? Who needed and used satellites?

Aside from satellite equipment manufacturers such as RCA and Hughes, there were also the new satellite carriers themselves. Costs of entry were formidable, far more than for terrestrial microwave networks. No ultimate assurance guaranteed success in the actual launch; in 1979, RCA lost a satellite valued at $50 million (including launch costs) (Brock, 1981: 265). As late as the mid-1970s, one authority explains, "there was speculation whether satellites could compete in the domestic communications market" (Morgan 1981: 26).

Four companies nevertheless quickly established a variety of satellite services utilizing two different satellite systems. RCA first undertook to lease channels on an orbiting Canadian satellite before launching its own a year or so later (1974). Western Union put up its own satellite in 1974. A subsidiary of Fairchild Industries, American Satellite, leased transponders on the Western Union system and provided its own earth stations; Southern Pacific Communications did the same. Services offered included television transmission, which rapidly emerged as a vital element in rapidly expanding "pay cable" systems, just as demand by television networks had helped to inspire the growth of microwave systems twenty five years before. Private line voice circuits and specialized services customized for individual companies also became available (Brock, 1981: 263). The latter, emphasized by American Satellite, employed earth stations at or near the customer's premises to hasten specialized applications. Dow Jones' transmission of the *Wall Street Journal* to regional printing plants makes use of such a configuration; other private networks were designed for Sperry Univac, UPI, AP and Muzak (Ibid.: 265). Close collaboration between satellite companies and prospective users marked the private network service field. Satellite Business Systems, a consortium comprised of IBM, Comsat and Aetna Life and Casualty, carefully worked out design plans with sixteen companies as part of its system development preparations (Ibid.: 276).

It was only at the close of the 1970s that satellite communications for specialized business purposes began to forge ahead by leaps and bounds. Not only had necessary trial experience been gained, but the technology of satellites had been vastly improved and advanced in step with microelectronics. Satellites, in contrast to terrestrial networks, today may allow all-digital transmission, with either point-to-point or point-to-multipoint capability, at a cost that is largely insensitive to distance. That is, once a signal has found its way from New York earth station to satellite, the cost of retransmitting it does not vary whether its destination is Chicago or San Francisco. For geographically dispersed companies with operating units spread across the nation, this feature is an obvious attraction when placed against the more distance-graded rates of terrestrial carriers. Once again, owing to large scale integration, digitized transmission and greater bandwidth (information carrying capacity), satellites may permit data traffic at speeds which rival the internal speed of computers themselves. Existing terrestrial Bell System circuits typically accommodate a data rate of up to 9,600 bits per second; with private line conditioning this can be raised to up to 500,000 bits per second. The SBS satellites, in contrast, permit a rate of 1.5 million bits per second, with a capacity for 6.3 million bps hoped for soon (Comsat *1980 Annual Report:* 17). For all these reasons, and because satellite circuit costs continue to drop with every advance in microelectronics, despite troublesome propagation delays encountered in transmitting signals up to the satellite and back, satellites are changing business communications.

The number of nonmilitary U.S. satellites in orbit grew from three in 1974 to nine in 1980; eighteen additional commercial satellites were authorized by the FCC in December of 1980. RCA Americom, a domestic satellite carrier, projects that while today four satellite carriers furnish a total of 168 transponders, a "realistic" picture for 1990 would be 984 transponders of which 590 would be allocated to "the entire gamut of business communications" (Inglis, 1981: 4–5). Earth stations have followed a similar trend, rising from five in 1974 to an estimated 4,000 in 1980; about 3,500 of this more recent total are receive-only stations for use by cable television systems and others, and priced far lower than sophisticated two-way data transmission facilities. Investment is also burgeoning, as the number of satellite carriers expands to seven. It is estimated that current committed expenditures to be made for satellites include: $198 million by GTE Satellite; $196 million by AT&T; $196 million by Southern Pacific; $190 million by Hughes; $100 million by Western Union; $200 million by RCA; and $400 million by SBS (Morgan, 1981: 20, 21, 24).

Businesses of all kinds are now rapidly incorporating satellite facilities and services into production, distribution and administration. American Satellite customers currently include General Dynamics, the *Wall Street Journal,* First Interstate Bancorp and 200 other corporate and governmental users. General Dynamics employs Amsat services for weekly payroll preparation, master scheduling and automated control functions, and centralized corporate data processing ("How Some Typical Customers Use SDX® Services," *American Satellite Capabilities and Service Profile,* 1981: 12). First Interstate Bancorp (formerly Western Bancorporation) relies on a Teller Item Processing System to interconnect 825 banking offices in eleven states. Amsat provides earth stations to the bank's three regional processing centers and to its control center in Los Angeles; the satellite circuits handle about 40 million of the 60 million bits transmitted over the network each day (Ibid.: 11). The *Wall Street Journal* and, now, the *New York Times*, make use of Amsat services for remote printing and regional distribution (Ibid.). Other users come from computer manufacturing, insurance, petrochemicals and pharmaceutical businesses (Ibid.: 11–12).

As the process of innovation and experimentation proceeds, new applications are emerging. The Holiday Inn hotel system established its own satellite teleconferencing network in 1980, using RCA Amcricom's Satcom 1 for two-way audio transmission. Today Holiday Inn has 170 satellite receiving stations installed at hotels throughout 40 states; by the end of 1983 the hotel chain intends to have 500 in operation ("More Satellites Crowding the Skies Mean More Useful Business Services," *Communications News* March 1981: 53). National advertisers are reportedly becoming interested in the potential of specifically targeted demographic groups reached on an ad hoc basis by means of specially designed programs distributed by satellite (Ibid.: 52). "Shopping channels" depending upon satellite signal transmission are being designed by Warner Amex Satellite Entertainment, and a group including Times-Mirror Satellite Entertainment and Comp-U-Card ("WASEC Keeps Perking Along, Plans To Add 3 More Channels," *Variety*, 2 September 1981: 38). Satellite carriage is thus a decisive factor in further concentration of consumer goods distribution in regional, national and, one presumes, international markets.

A further constraint on use of satellite circuits will be progressively weakened as satellites are linked to broadband cables and terrestrial microwave facilities, thereby effectively circumventing local loops of limited bandwidth operated by telephone companies. "The cable TV operators have the expensive coaxial cable system, which in many instances passes in front of a large number of offices, factories and computer facilities. On the other hand, the business satellite carriers are able to bring to a satellite earth station a vast amount of data, voice, electronic mail, telex and other business signals," explains Sidney Topol, chairman of Scientific-Atlanta (which produces equipment for both satellite and cable TV industries) ("More Satellites Crowding . . ." :52). "This has the makings of a strong interest in joint efforts by the two groups for an intra-city distribution system."

One such experiment is already in the offing, as SBS and Crocker Bank collaborate with Tymnet and Local Digital Distribution Company (a partnership of M/A Com and Aetna). Crocker will provide data for transmission between San Francisco and New York via SBS transponders. Cable TV lines will distribute the data traffic in New York; both cable and cellular radio—a now quickly emerging technique for local communications by radio frequency reuse—will be employed at the San Francisco end (Ibid.).

If the Domestic Satellite decision thus encouraged a tremendous surge in corporate use of satellites, and opened the way to further experimentation with the satellite medium, new institutional arrangements with unprecedented legal implications are now also developing. Companies have begun, under urging from satellite carriers hoping to cut back their risks, to *purchase* transponders for intracorporate communications. Having tested satellite use for some years, Dow Jones is purchasing two transponders on the Westar V satellite—to be used to tie together seventeen printing plants and other company operations (Thomas and Thomas 1981: 94); Westinghouse has options to buy transponders on Westar V or VI. Other companies are hastening to follow. Two advantages follow from private ownership by individual user companies. First, technical advance is expected to further increase the amount of data that can be squeezed into a given bandwidth—making ownership of transponders a good investment in future growth. Second, inflation and the volatile communications policy scene make ownership more likely as a favored path toward predictable corporate spending for communications (Ibid.). By what legal rights companies can come to own transponders making use of the spectrum, however, *without* exchanging their private control over this public resource for common carrier status, is, to say the least, debatable. The FCC Domestic Satellite decision, coupled with the agency's repeated willingness to authorize satellite launchings in its aftermath, thus have been of vital import and influence in the continuing integration of telematics into the structure of private production. This same result was also assisted, in a different way, by the Commission's authorization of "value-added carriers."

Specialized carriers had been licensed to build their own telecommunications networks to provide users with customized data-oriented circuits (more recently, as in the case of MCI, they have moved aggressively into the area of residential voice carriage). Satellite carriers further extended corporate users' freedom of choice. Distance insensitive and flexibly configured, satellites both brought down communications costs and increased service options for heavy users. The FCC decision to allow entry of value-added networks followed an identical logic.

Value-added carriers may furnish an unlimited variety of services. That which makes them a distinct class, however, is their reliance upon computers and software to add value to basic communications circuits which they lease from other, underlying carriers. Of especial importance in this regard have been "packet switched" networks. Unlike the conventional telephone system which is *circuit* switched to furnish a given user with exclusive use of a particular line (thereby largely prohibiting its employment by others, even though pauses and silences within voice conversations may occupy a quite substantial portion of use), *packet* switching relies on computers to arrange and store small groups of digitized data (packets) for transmission at very high speeds over the quickest or least expensive path available in a computer-controlled communications network. Packets which may be sent individually over different circuits are then reassembled in the correct order at the ultimate destination. Though utilizable for both voice and data traffic, packet switching is of particular relevance to data communications owing to its cost efficiency, the speed with which it transmits large volumes of data, the low error rates common to its use, and the comparatively brief connect time such use imposes. With appropriate software (computer instruction programs), packet switched value-added networks may allow terminals with disparate technical features to communicate with one another. They are thus well-suited to computer communications, having been adapted to an environment in which different computer manufacturers furnish users with otherwise incompatible equipment. Value-added networks, in other words, may add a necessary measure of rationality to circumstances arising from unstandardized systems marketed by diverse computer manufacturers. Packet switching was first developed at the U.S. Department of Defense, whose vast investment in computers presumably exerted pressures encouraging their intercommunication. The first commercial network authorized by the FCC was Packet Communications Inc. (43 FCC 2d 800) in 1973; PCI subsequently went out of business. Telenet was licensed in 1974 (46 FCC 2d 680), and five years later merged with GTE.

Entry of value-added networks onto the telecommunications scene constituted a dramatic change in the structure of the industry. Some such networks—the timesharing companies—perform computer processing on data transmitted from remote locations over common carrier circuits, or market database storage services. By the logic of the First Computer Inquiry, companies offering timesharing services were determined to be data processing firms making incidental use of telecommunications in operating their computer processing functions; hence, they remained unregulated. Others, such as Telenet and Tymnet, were packet-*switched* networks combining computers and communications in an integral service; therefore, the FCC categorized them as subject to common carrier regulation. Together these two classes of value-added carriers have existed in, and have themselves helped to enlarge, a "gray area" between regulated and untariffed service offerings (Sherman, 1981: 4). Thus by authorizing value-added networks the Commission quite consciously extended competitive user-oriented policies. Despite its recognition that entry of value-added carriers would markedly affect the structure of the communications industry, the FCC claimed to have "determined that entry should, nonetheless, be permitted because it would introduce new and improved means for meeting consumers' data transmission requirements in a manner not available from any generalized or specialized carrier" (Senate Subcommittee On Communications, *Hearings on Domestic Telecommunications Common Carrier Policies* 1977: 118). More tersely, the agency felt that such carriers had been licensed "to tailor existing carrier facilities to meet user needs" (Ibid.). Moreover, authorization of the new carriers added to pressures already building upon the established carriers—notably AT&T—to upgrade their networks to prepare for corporate data traffic.

We shall scrutinize other kinds of value-added networks in another section. Let us now briefly assess the nature of demand for *packet*-switched networks taken alone. Initially, benefits from packet switching were drawn by companies offering database and timesharing services (Wessler 1980: 373). Information retrieval services looking for more efficient data transmission and lacking extensive private network facilities were the earliest key users of the new carriers: Lockheed Information Systems, System Development Corporation, the National Library of Medicine (Ibid.). Major corporations, on the other hand, adopted packet networks more gradually. With dedicated networks already in place, it was only as need arose to replace or add to extant facilities that the largest private users embraced packet switching. In 1980, more than 500 public and private sector organizations employed packed-switched common carrier networks in the U.S. While, in 1976, the majority of GTE Telenet's subscribers were public computing or database service firms, a breakdown of the subscriber list for 1980 showed that a greatly increased number of private firms are using it for internal services. Approximately 30 percent were manufacturing and industrial companies; another 30 percent were companies in service industries such as banking, insurance and transportation. Public computing services accounted for less than 30 percent of the 1980 subscriber base, while government and public sector users took up the remaining 10 percent (Ibid.). To accommodate the greater traffic volume and to enhance transmission flexibility, GTE Telenet is introducing satellite facilities in addition to its ground network, and is pursuing advanced local distribution capabilities (Ibid.: 374).

Heavier user participation in packet switching, however, does not end with reliance upon public networks like Telenet. Large corporate users have also developed and implemented *private* packet networks between high traffic volume locations (Ibid.: 373). Although data are sparse on this subject, there can be no question that major users have become actively involved on an in-house basis. Manufacturers Hanover employs its own packet network and interconnects it with GTE Telenet—in turn allowing international connection to more than 30 countries (Ibid.: 374)—and facilitating private electronic mail to corporate users throughout the world ("Electronic Mail Cuts Bank's Phone Dependence," *Computerworld* 29 June 1981: 19). Howard Frank, president of a communications consulting firm, recommended recently that users whose monthly telecommunications bills exceed $100,000 should install their own packet networks (Hoard 1981: 13). Another prominent consultant, Howard Anderson, also predicted that the present period is one in which "(u)sers will start buying packet switchers themselves" ("At the Office with Howard Anderson: The View From the Sidelines," *Computerworld* 29 June 1981: 3). Once more, then, the outlines of a familiar pattern are growing clear, as the very biggest private users integrate and innovate a new telematics technology in-house to serve growing, specialized business needs.

BUSINESS USERS AND THE LEGISLATIVE BATTLE FOR POSITION

The dimensions of the conflict between corporate telematics users and American Telephone & Telegraph stand out more sharply upon consideration of the congressional context of their dispute. For, as the depth and importance of the debate over a national telecommunications policy became visible, the Legislature as well as the FCC was called in to participate.

In hearings in 1973 that focused on the communications industry, time was made available to the corporate communications manager of Montgomery Ward, Clinton W. Warkow (*Hearings on the Industrial Reorganization Act* Before A Subcommittee On Antitrust And Monopoly Of The Senate Committee On The Judiciary, 93rd Congress 1st Session On S. 1167, 30 and 31 July; 1, 2 August, 1973, Part 2, The Communications Industry, Washington, D.C.: USGPO, 1973). Warkow hoped to express "the communication user reaction and experience" to the evolving communications industry environment. He began, as would be typical also of users throughout succeeding hearings, by sketching the character of his organization's dependence on telecommunications. The story, by now, will be familiar. In 1972, Ward operated 2,155 sales outlets—including retail stores, catalog stores and sales agencies. Total sales in 1972 had been over two and a half billion dollars; total operating expenses had come to $654,412,000. Communications expenses had amounted to $18,072,548: a mere .0068 percent of sales—but a considerable 2.8 percent of operating expenses (Ibid.: 1127).

Warkow asserted that the established common carriers had been unable to provide the modems and private branch exchanges demanded by Ward. Having then purchased this equipment from independent "interconnect" companies, Ward had realized a 20.6 percent cost-saving over common carrier rates for comparable service. "There has been a strong willingness by most interconnect industries," Warkow stated, concerning the independent equipment manufacturers whose growth was spurred by the Carterfone verdict, "to modify their offerings to fit the demands of the user." This inviting flexibility was, he claimed, "much more readily available than that commonly shown by the common carrier" (Ibid.: 1130). Too, telephone company rates were pegged to the *value* of their service, Warkow complained, rather than to "a more precise 'cost of service' " (Ibid.: 1131). Asked whether "you went the route of privately owned equipment due to the inability of the telephone company economically to meet your communication requirements?" Warkow replied: "I would say this is one of the basic reasons, yes" (Ibid.). The actual availability of customized equipment was a related problem, for the telephone company tended to furnish extra unwanted features, costing extra, on equipment it supplied. The special financial advisor to the Subcommittee then generalized: "The gist of your testimony seems to be that by purchasing your own equipment you would immunize yourself from telephone company equipment rental rate increases." Warkow responded (Ibid.):

> Yes, that is one of the big selling features I felt in approaching our company originally with the idea of expanding this environment. A store operating manager in a Montgomery Ward store appreciates his ability to control a given expense and know exactly what it will be for a period of time. It is true he has to face the variable of maintenance service, but percentagewise that is a very small percent of his operating communication dollar.

Predictability, economy and control, required by companies expanding into a dispersed operating environment, were impeded by the monopolistic market power of the telephone company. For corporate users attempting to maximize each of these factors, "competition" was the only immediately available means to this end.

The trend toward greater private corporate control over telecommunications was made more visible during hearings in 1976 on a "Consumer Communications Reform Act"—basically, an AT&T-originated move to restore by legislative fiat what had been taken away by the several FCC decisions we have already assessed. The "Bell Bill," as it came quickly to be called, was the capstone in a vast public relations effort by AT&T to end the fledgling competition that had been allowed to erupt into domestic telecommunications. Users' sentiments were enunciated with clarity by William A. Saxton, president of DataComm User Incorporated—an affiliate of the International Data Group.

Saxton asserted that companies employed data communications, which the Bell Bill would have put back under the exclusive aegis of AT&T, primarily for three reasons: "savings, centralized control and fast turnaround" (*Hearings On Competition In The Telecommunications Industry* Before The Subcommittee On Communications Of The House Committee On Interstate And Foreign Commerce, 94th Congress 2d Session, 28, 29, 30 September, 1976, Serial No. 94-129, Washington, D.C.: USGPO, 1977: 727). Activity supporting these indispensable applications was concentrated, he said, at sites where computer systems had been installed. At the end of 1975, about $33 billion worth of general purpose computers (valued at original prices) had been installed at about 53,000 U.S. locations; slightly more than 14,000 of the latter were already utilizing data communications. Saxton predicted that around half of all 1976 expenditures on data communications (an estimated $4 billion total) would be made at organizations from among the top 100 Fortune companies. The top 500 corporate users, Saxton claimed, would account for 80 percent of expenditures for all data communications products and services (Ibid.). Such large private users had need for elaborate networks, crossing many state lines; it was this multilocational presence that made their need urgent. Because the trend in data processing was to put more computing power at remote sites "where the information is actually generated and/or used" (Ibid.: 728), "innovative and economical communications" had become ever more critical.

> telecommunications is the key to effective information management, which, in the remainder of this decade and in the 1980's, will spawn some unique benefits for the consuming public. . . .More sophisticated forms of information transfer will translate into lower prices for consumer services if user applications are freely allowed to evolve (Ibid.).

It must be underlined that Saxton was not referring to a "consuming public" comprised of individual persons. In his formulation as in others we shall find that the term "public" and the term "consumer" denote instead the large private companies which account for the bulk of expenditures on telematics. It was this actually private-sector "public" that, Saxton asserted, had been quite favorably impressed with the benefits flowing from competition in provision of communications circuits and terminal equipment. It was this elite group of the largest users which most looked forward to expansion of competition "primarily to save money, obtain more personal service and to have a real alternative to AT&T to keep everybody on their toes" (Ibid.).

The key unresolved question, Saxton circumspectly observed (Ibid.: 730), was "whether or not A.T.&T. can or will want to efficiently and economically provide all the datacommunications services that will be necessary to meet future needs." (By this point in his testimony, the issue of *whose* needs these were did not need to be clarified.) The telephone network had not been designed to meet data communications requirements. As a result, AT&T "is constantly adapting a system intended for another purpose to data communications." At the very least, this engendered "a monumental technical challenge"; but it begged the vital question of what forces would make such a technical transformation economically feasible. The Bell System, as we have seen, was doing quite nicely on its general voice service offerings; hybrid data services, while certainly a dynamic and fast-growing market, were still in absolute terms small when placed next to voice. If the Consumer Communications Reform Act was to become law and if, as a result, AT&T were to choose not to offer various vital data services, Saxton inquired, "who will if there are no competitors?" (Ibid.). "(T)he bottom line question that must be answered," according to Saxton, centered on users: "What will be the effect of lack of competition on the cost of a user's product?" (Ibid.). To answer it, he proposed that the Subcommittee should elicit testimony from "independent and objective groups representing the interests of all data communications users"—such as the Association of Data Communications Users or the International Communications Association, two prominent user groups that have been active in regulatory and legislative proceedings in subsequent years.

But at this point, Saxton offered an enigmatic claim. He said he was confident that these users, given a free environment in which to express themselves, would shed important light on the issues. "Please note my emphasis on the word 'free'," he added: "The Bell bill has already politically and emotionally charged the communications community to the point of discouraging individuals and their organizations from expressing their views" (Ibid.). He then closed his testimony and invited questions.

An ensuing exchange between Saxton, Congressman Lionel Van Deerlin and a previous witness provides a rare glimpse into some key private dimensions of the "public" policy process, and I therefore quote it in full (Ibid.: 731-732):

> VAN DEERLIN   I was somewhat surprised by the assertion just a moment ago about fear of expressing dissatisfaction. You found this among users of the telephone company services at what level? Who is afraid to speak up?

SAXTON I am talking about people in data communications management, or people who have responsibility for the data communications management function in their organizations, and the organizations themselves.

VAN DEERLIN Why should this be?

SAXTON It has been reported to me and members of my staff that certain pressures have been brought to bear on these people and their organizations for a plethora of reasons, most of which relate to business relationships between some of these organizations and A.T.&T. and its operating companies.

VAN DEERLIN You mean interlocking directorates and so forth?

SAXTON That is one example.

VAN DEERLIN I see our previous witness behind you nodding and gesticulating.

FEINER Right. (Robert L. Feiner, President, Phonetele Inc., a small terminal equipment firm)

VAN DEERLIN Do you have some additional information on this you might share with us?

FEINER Mr. Chairman, by way of example, a couple of months ago I submitted a petition to the New York Public Service Commission to cause New York Telephone Co. to provide certain data (New York Telephone Company is an operating company in the Bell System—DS). As a matter of what I thought to be pure courtesy, I included on the distribution list among legislators the members of the International Communications Association. They are major company telecommunication staff people who have a vested interest in what we are trying to do, representing major corporations doing business in New York. It seems that I created a storm.

    I got a call from the president of the association; he was having certain complaints. I got four hostile letters. One of them I could not check back as the company is not a public company, but three of those companies had interlocking directorships with either A.T.&T. or New York Telephone Co.

	And, as a matter of course, the associations will not involve themselves in any regulatory proceedings or any of these arguments, will not take a position, pass a resolution, or do anything because they are afraid of all the pressures and will not get involved. (ICA amended its statutes in 1979 to allow policymaking intervention and created a Government and Regulatory Affairs Liaison Committee for this purpose - DS).
SAXTON	Let me add, Mr. Chairman, these associations - and the gentleman is quite correct - generally, by charter, are forbidden to take a position on public issues of this type. However, they are available to lend the strength and the profile of their organization to studies on vital issues in this area. I am certainly in no position to make commitments for the Association of Data Communications Users in view of the fact that I am just their advisor. It has been expressed to me, however, that they are more than willing to participate or even solely undertake the type of studies that would be necessary for members of the committee and other industry observers to help you make the kinds of judgments that are before you. But the positions of the individual members of these associations are affected more by their organizations than they are by the associations of which they are members. They wear two hats and, in certain instances, both of their hats have been squashed.

Congressman Van Deerlin, apparently recognizing that he had somewhat carelessly opened up an exceedingly delicate field of inquiry, then interjected—

VAN DEERLIN	At the least we are not short of witnesses, so I guess we will—
SAXTON	Let me also make one concluding point, Mr. Chairman. In the face of some data communications management professionals and their organizations not being able to take a position on this matter, there are some who have already taken steps to clear the way for their being involved in this; some of the leading users and more forward-looking organizations recognize the potential impact on their overheads and they have authorized some of their people to come into the public view with their opinions. So it is not a totally desolate situation.

Of several remarkable aspects of this exchange, I single out two: the apparent power of AT&T to shape domestic telecommunications policy through private means, and the gathering response from major corporate users that this exercise of corporate power had clearly provoked. A 1978 Senate staff study concluded that numerous director interlocks "could supply convenient conduits for possible private resolution of the public debate between monopoly and competition in the telecommunications industry" (*Interlocking Directorates Among The Major U.S. Corporations*, Staff Study for The Committee On Governmental Affairs, U.S. Senate, 95th Congress 2d Session, Washington, D.C., USGPO: January 1978, 102). More recently a study of 100 front-rank U.S. companies found that, within this elite group, AT&T had not less than 39 direct and 938 indirect board interlocks—more than any other firm in the group, which comprised companies in every major economic sector (*Structure Of Corporate Concentration*, Staff Study for The Committee On Governmental Affairs, 2 Vols., U.S. Senate 96th Congress 2d Session, Washington, D.C., USGPO: December 1980, I, 25). Standing out in the above-quoted exchange is evidence that company insiders fully believed the Bell System made use of such interlocks to support its corporate objectives. From their perspective, perhaps, as from that of independent analysts, the Bell System has long been known as "a unified power complex that is capable of being mobilized in time of need" (Herman, 1981: 227).

This was clearly such a "time of need." It is imperative that we be quite specific about this point. The Bell System had suffered a series of reverses at the hands of the FCC—none fatal, to be sure, and taken by itself, no one of which immediately threatened AT&T with imminent disaster. These reverses signaled cumulatively, however, that the Bell System's unique ability to influence and guide the regulations to which it would then be subjected by the FCC, was being effectively contested. Large corporate users did not disdain AT&T, nor did they at the time wish to see it dismembered. Rather, they demanded only that Bell furnish up-to-date and economical data communications services and equipment to meet their needs, and that they be allowed to employ the public switched network dominated by AT&T for specialized corporate purposes. AT&T on the other hand resisted such changes not only because it was making out quite nicely on voice services but, ultimately, because the market power of the Bell System has been closely tied to its single-minded dedication to a single public switched network under its control. Users, from the Bell perspective, were attempting to impose a whole series of limits and constraints on that unified control over the network: through liberalization of terminal equipment and interconnection rules, through demands for unregulated data processing services making "incidental" but nonetheless profoundly vital use of telecommunications circuits, and through a number of even more basic demands, first voiced in the First Computer Inquiry, like the request by major users that resale and shared use of common carrier-provided circuits be permitted. Should such far-reaching changes be accepted, the market in which the Bell System operated would be radically enlarged while its own power to direct that market would be correspondingly truncated.

AT&T therefore used every means at its disposal to alter the evident course of domestic telecommunications policy. We have briefly glimpsed a rare instance of the giant telephone company flexing it muscles in the private sphere. The Consumer Communications Reform Act, which was co-sponsored by 175 representatives and sixteen senators at its introduction, was the company's official public response. The culmination of a massive public relations and lobbying effort, the Bell bill was supported in public testimony before the House Communications Subcommittee by no less a figure than John D. deButts, then chairman of the board of American Telephone & Telegraph.

Despite formidable initial support, the Bell bill was defeated. How may we account suitably for this long string of reverses suffered by AT&T, ranging from the Above 890 decision and Carterfone, through Specialized Common Carrier and Domsat, up to and including the decisive defeat of the Bell bill? According to a recent analyst of corporate structure, "(i)t is not possible to alter the structure of power without the mobilization of equivalent or greater power" (Herman, 1981: 294). We must therefore identify an entity exercising power equivalent or greater to that held by the Bell System itself. No one company can possibly fit the bill—AT&T is bigger than any other. It can only be that we are looking for a group of companies large enough to set aside the telephone giant's own preferences. The only force equal to such a task is comprised of the very group of corporate telecommunications users that, as we have seen, has for decades both demanded new integrated telematics equipment and services, and been increasingly dissatisfied at the response offered by Bell. This group of big business users is the irresistible force that contests against the immovable object that is AT&T.

Their struggle must not be reduced to any simple opposition. Many of the largest users are tied to the Bell System through a variety of means—director interlocks, for instance, or procurement practices. Nor should the conflict be anthropomorphized; for this is not a flesh-and-blood battle in which one party to win must finish off the other. It is, rather, a struggle for command over the evolving direction and shape of the national telecommunications infrastructure in which both parties—AT&T and corporate users—seek relatively greater powers. It is, as I have said, a war for position. But if the Bell System has been able to act as a "unified power complex," this was far less true for business users.

Users needed to organize themselves into a more cohesive force so as not to be whipsawed one-by-one by the Bell System. In part, however, development of user organizations to represent them in policy debates was an outgrowth of technical need. The postwar period, we have seen, witnessed vast enlargement of private networks among the largest corporations. Finding optimal solutions to innumerable technical problems inherent in owning, or leasing and managing, fast-growing private networks, became more crucial in like proportion to the expansion of private networks themselves. Even by the later 1940s, users had started to band together to share technical and operational information about how best to run their telecommunications facilities. Today this process of organization is very far advanced.

User groups have taken a number of forms. On one hand, they may be formed specifically around particular technical modes, such as the International

Electronic Facsimile Users Association, formed in 1979 as a nonprofit trade association representing the user of facsimile equipment ("Incorporate Fax Users Association," *Communications News* March 1981: 20). Over 60 companies, both in the U.S. and abroad, have become members of this group. Also of this type are the International Association of Satellite Users and the Association of Data Communications Users. On the other hand, user groups may emerge to serve the needs of discrete sectors of industry. The Wall Street Communications Association, for instance, concerns itself both with voice and data communications, but includes members whose major function is that of security brokers and dealers ("Wall Street Group Elects New Officers," *Communications News* September 1981: 15). Or, as in the case of banking and energy, communications organizations serving specialized user needs may evolve from extant trade associations: the American Bankers Association's Communications Group, and the American Petroleum Institute's Central Committee on Communications Facilities are of this type.

A third axis along which user groups may cluster stems from geography; companies (and, often, public sector organizations as well) in a given state or region may well face particular local and regional issues of both technical and regulatory significance. The Iowa Telecommunications User Group was organized to "provide a means whereby users of telecommunications facilities and services can exchange information, experiences and concepts to the mutual benefit of the individual members and their companies, and to encourage technological research and development in the field of telecommunications by means of seminars, conferences, newsletters and special reports" ("Organize Group for Iowa Users of Telecom Services," *Communications News* June 1981: 21). Membership consists of directors of telecommunications of corporations and other organizations headquartered, or maintaining branches, in Iowa. Meetings are held monthly. Other states and regions maintain analogous groups.

More encompassing are the Midwestern Telecommunications Users and—though it has of late become more fully a national presence—the Tele-Communications Association, which represents some 1,200 members from 500 companies ("Mike Woody and Dan Grove Tell of TCA Plans and Progress," *Communications News* September 1981: 52–57). Still larger, and probably the broadest-based user group in the United States today, is the International Communications Association. ICA represents 470 organizations; to qualify for membership, an organization must spend in the vicinity of $1 million annually on telecommunications, must operate multi-city offices, and must not be predominantly engaged in the production, sale or rental of communications services or equipment. ICA thus speaks for a broad group including some of the largest corporate users—and has done so since 1948, when it was formed by representatives of a select group including Sears, Dupont, GE, GM and U.S. Steel. Informational exchanges conducted through various media and mechanisms facilitating liaison with educators, common carriers, telecommunications suppliers and government and regulatory agencies, are prominent among ICA's activities. Industries represented on ICA are diverse, embracing energy, foodstuffs, aerospace, pharmaceuticals, banking, computer equipment and services, chemicals, automobiles, and publishing. According to its filing in a recent FCC proceeding, the telecommunications expenditures of ICA members amount to six billion dollars annually—a significant fraction of total telecommunications industry revenue (Comments of International Communications Association, Before the Federal Communications Commission, "In the Matter of Regulatory Policies Concerning Resale and Shared Use," CC Docket No. 80-176, 7 August 1980: 5).

User organization has proceeded apace not just with technical necessity but, increasingly, with regulatory policy. In Part Two there will be need to outline the *international* activities of various user groups in some detail. In the domestic sphere, it is as well to mention that user organization is dynamic, not static; the most recent legislative attempt to rewrite the 1934 Communications Act has apparently provoked extant user groups to consider forming a "user coalition." "It is time," asserts Tele-Communications Association president Mike Woody, "to come together in a unified voice saying this is how we, as users, perceive the needs" ("Mike Woody and Dan Grove Tell of TCA Plans and Progress, 1981: 52). User groups have become pivotal means to gain political and policy influence, locally, regionally, nationally and internationally. In this regard they are not different from the trade associations from which they sometimes emerge, in that they comprise a means for inter-industry, intra-industry and industry-government coordination (Cf. The classic account, Brady, 1943; more recently, Herman, 1981: 212–214). Let us briefly now scrutinize the nature of users' intervention in the legislative realm in the period since the Bell bill.

In Senate hearings during 1977, for example, legislators heard testimony from Robert Barnwell, corporate manager of office services for Fiberboard Corporation and president at that time of the Tele-Communications Association (*Hearings On Domestic Telecommunications Common Carrier Policies*, Before The Subcommittee On Communications Of The Committee On Commerce, Science, And Transportation, U.S. Senate, 95th Congress 1st Session, 21, 22 March, 1977, Two Parts. Serial No. 95-42. Washington, D.C., USGPO: 1977). Founded in 1961, TCA was by 1977 the largest regional organization of its kind, serving the Californian southwest. Members came from many industries: airlines, trucking, railroads, aerospace, news media, banking, insurance, lumber, retail, medical, etc.—as well as from educational institutions and government agencies (Ibid., Part II: 1045). The basic objectives of TCA were to provide continuing education for members through formal and informal programs, to encourage improvements in telecommunications technology, and to provide liaison with educators, government agencies, carriers and other telecommunications suppliers.

His introductory remarks finished, Barnwell went on to outline the key objectives of corporate telecommunications managers, such as those represented through TCA. These goals included (Ibid.): improved telecommunications system capabilities both within corporations and between them and their customers;

provision of "innovative, cost-effective communications systems that enhance company operations through increased productivity" and that "conserve corporate and natural resources"; application of new technology "to upgrade existing corporate telecommunications systems to reduce cost and improve service"; and, finally, pursuit of "regulatory and legislative policies that foster a climate conducive to technological innovation in the telecommunications industry."

This was no abstract concern for "new technology" and its inherent merits. It was, instead, a suggestion that *specific* new techniques be permitted to develop and that this development take place under the direction of corporate user needs. More particularly, Barnwell stressed that industry wished "to further merge voice, data, message, and video communications into a single viable medium" (Ibid.: 1047). To achieve this goal both more bandwidth and a capacity for digital transmission were necessary. Satellites, packet switched networks and fiber optics were identified by Barnwell as comprising a positive presence on the telecommunications scene. The day-to-day business of industry required "the movement of huge volumes of paper as well as frequent personal travel," and such indispensable communications were growing costlier (Ibid.: 1046). "We believe," declared Barnwell, that "telecommunications can be the means that will allow the Nation to reduce these costs" (Ibid.)—as if it were somehow transparently obvious that the nation as a whole would indeed obtain "lower prices for goods and services," merely because telecommunications were significantly improved. Again, Barnwell believed that implementation of technology that would "significantly reduce the cost of shorthaul facilities" was "necessary" (Ibid.). At bottom, then, Barnwell was arguing for the classic user agenda: digitized circuits facilitating data traffic at lower cost for major corporate users.

He was followed and largely supported by the senior vice president of the Securities Industry Automation Corporation (SIAC)—a jointly-owned subsidiary of the New York and American Stock Exchanges formed in 1972 "to assist those exchanges in improving the securities marketplace environment for the investing public, through innovative and cost-effective data processing and communications systems" (Ibid., Statement of Vincent P. Moore, Jr.: 1048). Apart from numerous clearance and settlement responsibilities, SIAC was operating manager of SECTOR—the Securities Telecommunications Organization—a not-for-profit industry communications service with 290 member-firm users, employing over 1½ million miles of carrier circuits supplied by five terrestrial and three satellite carriers.

Moore placed an unusual stress on the history of his industry, stating that from its 18th century beginnings, the securities industry had "always placed a paramount value on the interchange of information" (Ibid.: 1049). For this reason, it had tended to place an equally prominent emphasis on new technology. The introduction of the stock ticker in 1867, the use of telephones on the floor of the exchange in 1878, scarcely two years after its invention, application of computers in 1961—"all hallmark an early and active role in the application of information systems." Information exchange was "basic to the securities market process" in the form of bids, offers, last-sale prices and confirmations (Ibid.).

Freedom to pursue further innovation across the broad field of emergent telematics techniques and services had once more become vital. And cost effective communications—"no small matter to our industry"—had become increasingly urgent. An annual survey of 400 New York Stock Exchange firms doing public business showed that they had spent more than $500 million on information movement in 1976 (the figure included private line circuits, message telephone service, quotation tickers and, significantly, postal service). Collectively these costs formed the third largest expense item for member firms, following manpower and interest expenses (Ibid.: 1050). As pressures developed for a national market system to emerge, Moore asserted, the cost of telecommunications was bound to increase dramatically.

Because it was both technically feasible and economically attractive, SIAC's vision of the future centered on transformation of the telecommunications infrastructure. Digital systems would "avoid the expense of maintaining separate voice and data networks"; satellite services structured around company or industry networks of earth stations would propel further growth; high speed facsimile networks permitting communicating word processors and other advances would bite deeply into current postal expenditures. Two regulatory policy requirements, however, were essential to support and sustain this "forward direction" (Ibid.). The first was that selfsame "freedom to innovate" that characterized the entire history of the securities industry. The second was "a broad diversity of suppliers—to avoid an undue concentration of power in one entity, which, in turn, could adversely affect an industry as communication intensive as ours," and, at the individual firm level, such diversity alone would "allow users a choice" (Ibid.).

Real competition between telecommunications suppliers was necessary for two related reasons. First, the newly authorized specialized carriers had "excelled" in "marketing to the particular communications needs of industrial segments" (Ibid.: 1055). That is, their smaller size permitted the new carriers to focus intensively on specific categories of users and their particular needs. They had already, in fact, "devoted substantial efforts to the tailoring of their services to the needs of particular customer applications." Their facilities, too, had shown high quality and "consistently responsive service offered at attractive prices" (Ibid.). Second, their efforts had forced a grim appreciation upon the Bell System; "in response, Bell has revitalized its marketing effort and directed that effort along the line of specific industry requirements" (Ibid.). With its heavy expenditures for communications, SIAC had therefore chosen to structure its communications service, SECTOR, as a matter of explicit policy, "to permit the participating member firms to select from as broad a number of carriers as possible" (Ibid.: 1050). Moore was able to report that the community of member firms "has responded very favorably within this policy framework" (Ibid.).

Robert Capone, vice president and director of systems and data processing for J. C. Penney Company, followed Moore and agreed wholeheartedly with his policy stance. After delineating in specific terms—number of terminals, number of catalog stores in need of interconnection to central computers—his company's "dependence on telecommunications for the effective operation of our business" (Ibid.: 1057–1058), Capone led up to a position statement. "We believe, very strongly, that competition has been the key element in allowing us to move ahead as rapidly as we have" (Ibid.: 1058). Penney purposefully chose terminals from a variety of suppliers; likewise, it preferred to obtain circuits not only from Bell but also from Southern Pacific Communications and MCI. Modems were found from a similarly diverse field of manufacturers. In short, Capone claimed, "We are constantly exploring and looking for new sources of supply in all of these areas" (Ibid.). Network design and terminal development for the Penney telecommunications systems grew out of intensive cooperation between various manufacturers and "our own people" (Ibid.: 1059). Future plans called for analogous expansion and integration of new techniques for word processing and electronic mail as had been mentioned by SIAC and TCA. Capone forcefully came out in favor of a regulatory climate that would stimulate technical innovations tailor-made to specific needs felt at Penney: "In these days of rapidly changing and improving technology it is imperative that our option (sic) continue to remain open. Our goal is to make sure that our telecommunications systems are the most effective and efficient available. We must, therefore, be allowed to choose those suppliers and systems that best meet our needs" (Ibid.: 1058). Akin to its trade association, the National Retail Merchants Association, Penney therefore found that any threat to competition "in the extremely small segment of the telecommunications market where it now exists"—a threat such as had been formalized in the Bell bill—would be "a serious disservice to the American consumer" (Ibid.). As had become usual, the word "consumer" in such a context denoted large private telecommunications users.

Since AT&T kicked the telecommunications policy ball into Congress with its Consumer Communication Reform Act, conflict over the shaping of a national telecommunications policy has widened, deepened and sharpened. Successive attempts to rewrite the 1934 Communications Act, in addition to their many other facets (including, at different stages in the process, broadcasting and international communications as well), have become a battlefield upon which, largely ignored by analysts, Bell and user groups have tilted head on. Users have had one victory thus far: since the Bell bill died all major policymaking in the telecommunications field has tended to be argued and justified in light of the prime user standard, *competition* in provision of equipment and services. What has not been clarifed, on the other hand, are the nature and extent of whatever competition is engendered.

Of notable import in this painfully complicated process has been the prolonged user-backed attempt to secure a host of unusual safeguards to protect the "public"—that is, the corporate user "public"—from ill-treatment at the hands of the Bell System. Through their lip service to the doctrine of competition, we find the curious and ironic spectacle of corporations that are oligopolies and monopolies in their own markets justifying their *opposition* to monopoly market power in telecommunications by reference to the rhetoric of eighteenth century entrepreneurship. Through their concern to prevent the Bell System from dominating supply of the essential telecommunications upon which they have come to depend, we see users demanding price competition and public participation in regulatory proceedings, as well as strengthening of regulatory authority to defend this special sort of public interest—all of which they often find less palatable when granted to individual citizens in the name of a general public interest.

By 1978, representing 300 of the Fortune 500 manufacturing concerns, together with many other non-manufacturing companies, the membership of the International Communications Association accounted for approximately 13 percent of the gross annual revenues of the Bell System (*The Communications Act Of 1978*, Vol. II—Part 2, Hearings Before The Subcommittee On Communications Of The Committee On Interstate And Foreign Commerce, House of Representatives, 95th Congress 2d Session On H. R. 13015, 9, 10, 14, 15, 16 August, 1978, Serial No. 95-196, Washington, D.C., USGPO: 1979, 158). "We are in a position to present this general consensus of our views upon the important basic questions," declared the chairman of ICA's regulatory liaison committee, Robert Bennis (Ibid.: 164). It was "critical" to user interests that any new legislation "promote the availability of a wide variety of communications services and equipments," he explained. Today's sophisticated user, he went on, "needs a broad flexibility, consistent with the foregoing, to develop the common carrier or non-common carrier services and equipment mixture which is most suitable to meeting its own particular communications requirements" (Ibid.). Such flexibility, he maintained, should be encouraged under any new legislation and, he surmised, "we believe this to be the fundamental underpinning of the legislation we are discussing" (Ibid.: 165–166). And, Bennis insisted (Ibid.: 166): "To the extent that such communications services and equipments are to be provided in a regulatory environment, it is also critical that the legislation adopted be structured in a fashion that encourages the speedy introduction of technological innovation, and that there be a practical and realistic opportunity for effective participation by the public in the regulatory process." ICA endorsed the general rhetorical emphasis of the bill on maximizing competition while minimizing regulatory intrusion (Ibid.: 167).

Yet Bennis announced that the bill's attempt to place maximum reliance on marketplace forces (Section 311), and to mandate federal regulation only where marketplace forces were deficient (Section 101), "cannot be placed in a vacuum and expected to perform their intended function" (Ibid.: 167). More precision was needed, for many provisions in the bill lacked "guidelines and definitions." Thus, "serious uncertainties" were inherent in its present language, and its provisions as they stood "could be construed in ways clearly detrimental to user interests'' (Ibid.). In brief, while ICA certainly supported the goal of competition, "(t)he transition from here to there...requires expert management and most careful planning" (Ibid.: 159).

ICA had apparently studied the prospective rewrite bill meticulously with an eye to injectiong *more opportunity* for "public" intervention in policymaking and *more rigorous* regulatory safeguards. As major users with a vested interest in telecommunications, Bennis asserted, ICA members "have a particular concern that there be a full opportunity to participate in the consideration of carrier filings" (Ibid.: 168). They had been alarmed to find, however, that provisions in the bill governing such filings might "drastically reduce" or even "eliminate" existing user opportunities to be heard (Ibid.). For instance, the bill would reduce by half (from 90 to 45 days) the advance public notice period for new tariff filings—even for the most comprehensive overhaul of current tariff structure. Such a reduction would severely limit user intervention, as it would bar meaningful review of voluminous and complicated cost and revenue data, impede identification of potential problems, obstruct formulation of a legally coherent user position, and place radical constraints on the ability of a large organization like ICA to poll its membership in time to reach a position platform (Ibid.: 168–169). The bill was imprecise and unpredictable and tended to sustain formalistic constructions over substantive safeguards. A rate would become "equitable and lawful," for example, merely if the oversight agency had not concluded a hearing and reached its decision within nine months (Section 314 (b) (3)). ICA well understood "the frustrations experienced in many common carrier rate making proceedings," and shared the view that improvement in disposition of such matters was essential. "But, at the same time it is also critical that the issues in these proceedings be disposed of *on their merits* rather than by the expiration of a predetermined time period" (Ibid.: 171), and, for this reason, the practical impact of a legislated nine-month schedule might "cripple user, or even agency, participation in the hearing process" (Ibid.). The carrier would develop "a virtually irresistable (sic) incentive to initiate delays" (Ibid.: 172), in contrast, and therefore no such schedule should be mandated.

Several other serious problems were evident. The language employed to create the new Communications Regulatory Commission (CRC) to replace the FCC, granted it potential power to exempt carriers and services totally from regulation and, in ICA's view, "the new agency should not be left with such a broad mandate" (Ibid.: 181). As Bennis reasoned, "the public is not properly served by vesting virtually unbridled and undirected discretion in five appointed officials" (Ibid.: 181–182). Clarification was required. While ICA wished to restrict the CRC's capacity for action where it might lead to greater freedom for the carriers, in other respects the group cautioned that the bill's emphasis on "streamlining and minimizing regulation may well have underestimated the complex responsibilities which would be left with the new agency" (Ibid.: 185). Effective implementation of common carrier regulation would more likely follow if at least some commissioners were "required to spend essentially full time upon such matters" instead of diffusing their effort across the vast field of communications-related matters (Ibid.: 186). Users were on the offensive and pushing for a regulatory agency devoting itself actively to a well-demarcated domain instead of the usual alternative—a sharply restrained agency with extensive responsibilities over huge territorial reaches, lacking both budget and manpower for effective oversight. Users in effect demanded more political democracy—for them.

It was not clear what effect the bill would have on currently unregulated services—those which, in ICA's view, would serve user interests best by remaining unregulated (Ibid.: 173–174). Explicit language was necessary to clarify the degree of jurisdiction of the CRC over telecommunications services, particularly in regard to adequate and precise differentiation between "competitive" and therefore unregulated services, and "noncompetitive" and thus regulated offerings (Ibid.: 175–176). The key to it all was the construction placed upon "competition." A substantial measure of competition might exist in some geographic locations and not in others, declared Bennis; or within certain subcategories of service but not in all; or at one time of day but not at another. Fundamentally, "to assume that all users are protected because of the availability of some degree of competitive alternative is an assumption we think premature" (Ibid.: 177). The National Retail Merchants Association agreed, that the "competitive principle" at the heart of the bill, should be "carefully defined and gradually implemented to recognize that competition in the telecommunications industry is today far from complete, and that the only truly viable competition from the consumers' standpoint is that which promotes *lower prices* as well as diversity of supply" (Ibid.: 1061). NRMA went so far as to castigate the "oligopolistic" market structure of the telecommunications industry (Ibid.: 1066), and concurred with ICA that the new federal regulatory agency should be vested with "sufficient authority to react to gradual development of full and free competition. . .and embody full procedural protections for telecommunications users" (Ibid.: 1073).

Telecommunications has simply grown too important to business users to be deregulated in some cavalier fashion, and without scrupulous attention to substantive "public" safeguards. In various succeeding versions of the Communications Act, users have therefore consistently argued in favor of real competition in telecommunications services—for under continued AT&T monopoly, the freedom of choice gained by any given corporate user to employ telecommunications as desired will, users collectively feel, be substantially diminished. We thus find the National Retail Merchants Association protesting that in a 1980 version of the bill, the mandated transition to a deregulated environment should "not occur in ways that inhibit competition" (*Telecommunications Act Of 1980*, Hearings Before The Subcommittee On Monopolies And Commercial Law Of The Committee On The Judiciary, House Of Representatives, 96th Congress 2d Session On H.R. 6121, 9 and 16 September, 1980, Serial No. 69, Washington, D.C., USGPO: 1981, 751). A more recent rewrite attempt, S. 898, was similarly attacked by the

ICA for assuming that by 1982 there will be enough competition to support deregulation of key services (Phil Hirsch, "User Faults New Communications Act Rewrite," 1981: 8). The leader of the Tele-Communications Association, Mike Woody, testified to the still only nascent competition in telecommunications, asserting that "the only protection from abusive pricing and related practices is rigorous regulation. . .to be phased out as competition becomes more effective" ("TCA Advocates Changes," 1981: 57). The current House variant of the bill—H.R. 5158—incorporates an identical user-backed standard.

Meanwhile the dynamic integration of computers and telecommunications continues, driven ever forward by business user demand for such merged services and by a favorable regulatory and policymaking environment. Let us turn to the emerging new order in domestic telematics.

THE EMERGING RESALE ENVIRONMENT

In the wake of the unsuccessful Consumer Communication Reform Act—the Bell Bill—the conflict between AT&T and the large business user community has swirled into sharp focus both at the Federal Communications Commission and, more spectacularly, in the just-settled antitrust case brought by the U.S. Department of Justice against the telephone company.

The Commission, for example, specifically underscored that passage of the Bell Bill "would seriously disserve the public interest by limiting important consumer rights" (*Hearings On Domestic Telecommunications Common Carrier Policies*, Before The Subcommittee On Communications Of The Committee On Commerce, Science, and Transportation, U.S. Senate 95th Congress 1st Session, 21, 22 March 1977. Two Parts. Serial No. 95-42. Washington, D.C., USGPO, 1977: I, 71). Passage, moreover, would constrain effective regulatory action "to protect consumers against discriminatory rates and practices." Finally, passage of the bill would undercut the public interest "by delegating to the unreviewed discretion of the telephone industry fundamental social and political decisions" (Ibid.).

Elaborating, the Commission asserted that by the late 1960s it "was being confronted with user complaints about the unavailability of facilities and terminals which could meet their communications needs, while, concurrently, new companies were emerging who wanted an opportunity to serve those unsatisfied communications needs" (Ibid.: 73). It was becoming "obvious" that a fundamental issue loomed before the Commission (Ibid.):

> Would the communications needs of the American consumer best be fulfilled by entrusting exclusive responsibility for the provision of all communications services to the traditional carriers (AT&T, the Independent Telephone

companies and Western Union); or, would the public interest be better served by permitting the provision of specialized facilities or terminal equipment by any individual or company ready, willing, and able to meet consumer demands for service?

Thus the Commission equated the public interest as a whole with the restricted private interest of the corporate user public in response to the repeated and concerted intervention chronicled above. An "ancillary—but equally fundamental—issue" betrayed the same realignment (Ibid.):

> Should the American consumer have the freedom and right to select among various potential suppliers of communications services; or, should his options and alternatives be limited to those services which his or her local telephone company should choose to provide?

Phrased carefully in terms of animate persons, the question itself was of unlikely relevance without massive involvement by corporate users in FCC deliberations. The very diversity of telematics applications within the sphere of corporate production brought a third related question to the forefront of agency thinking (Ibid.):

> Was the basic economic rationale for having a monopolistic structure still applicable to all segments of the communications market? Or, would competition in those changed markets be both feasible and likely to produce benefits . . . ?

Would corporate user demands, in short, be permitted to alter—in ways as yet largely unforeseen—the essential structure of the American telecommunications industry?

To this basic question the Commission answered with a resounding "yes!": acting as lead agency in the computerization of society, the FCC continued, in proceeding after proceeding, to articulate and respond to an evolving user agenda.

And, as the Commission went about its work, a second, omnipresent threat to the structure of the telecommunications industry piled another layer of uncertainty onto the process of policymaking. Would AT&T—*could* AT&T—be broken up by the antitrust suit brought by the Justice Department on 20 November 1974?

The ultimate shape of the dynamic telecommunications industry is still far from clear. As the range of participants in the great telecommunications policy debate continues to expand, and as the various stakes in the outcome mount with growing recognition of the elemental role of telecommunications in the U.S. economy, it becomes virtually impossible to predict that outcome. Broad characterization of underlying trends and policies may nevertheless silhouette at least the general outlines of an emerging order for domestic telematics. Let us turn now, therefore, to some of the changes that are today coming so swiftly to telecommunications, partly as a result of the spiraling conflict between business users and AT&T.

Resale and Sharing

The FCC's decision to authorize value-added networks was momentous far beyond the existence of a few new carriers—PCI, Graphnet and Telenet. Value added networks that applied switching equipment to private line circuits leased from AT&T, the underlying carrier, were reselling these circuits in competition with AT&T itself. As the Commission put it, resale is "an activity, wherein one entity subscribes to the communications services and facilities of another entity and then reoffers communications service and facilities to the public (with or without 'adding value') for profit" (60 FCC 2d 271).

By custom, common carrier tariffs sharply limited resale of private line facilities and services—other than for a few special customers. AT&T, prohibiting resale as a matter of general policy, thus voluntarily permitted resale of certain facilities by its old—and unthreatening—rival, Western Union. Closely related is "sharing," a nonprofit arrangement where several users collectively acquire and use communications facilities provided by an underlying carrier, with each user paying the costs of its pro rata usage of the circuits (60 FCC 2d 263). AT&T had again voluntarily permitted sharing for a very few specific groups: Aeronautical Radio, Inc. (ARINC), for instance, which represents airline companies in communications matters, was recognized as a legitimate "single customer" for AT&T private lines—even though ARINC itself allocated these circuits as needs dictated to members—in 1937 (Aeronautical Radio Inc. v. AT&T, 4 FCC 155). Subsequent requests to AT&T by other prospective user groups, seeking the requisite single customer status to permit them to innovate sharing of private lines, suggested an increasing demand (60 FCC 2d 267).

The inquiry into resale and sharing opened in July 1974, and was needed to furnish answers to two basic questions. First, "what constraints may carriers owning and operating transmission facilities lawfully impose upon the use which customers make of the service provided by these carriers?" (60 FCC 2d 268). In its deliberations, the FCC specified that an answer to this question would depend "not only upon an analysis of legal precedents"—which, presumably, showed that resale and sharing had been strictly circumscribed—but also "upon economic and policy considerations today in light of the growing demand for specialized, customized services and the technological developments in the communications field" (Ibid.). Second, should resale and sharing be more widely authorized, to what degree should the FCC regulate the new networks that would ensue? (Ibid.).

It was evident at once through their filings that the proceeding was relevant to the interests of groups far beyond the new value-added carriers; a partial listing of comments in the docket includes those of Aeronautical Radio, the American Newspaper Publishers Association, the American Petroleum Institute, Associated

Press, Bank Wire, Citicorp, Dow Jones, National Association of Manufacturers, National Retail Merchants Association, and the Securities Industry Automation Committee.

In July 1976, the Commission resolved to permit unlimited resale and sharing of common carrier private line facilities and services (60 FCC 2d 261). Entities that resold communications services would henceforth be regulated as common carriers by the FCC. However, to encourage open entry into the resale services market, the Commission declined to require resale carriers to show cause of public need for their prospective service as a condition of certification. Sharing arrangements would not be regulated. The decision to expand resale and sharing was expected, the Commission stated, to bring about "public benefits." Notably, it was held to facilitate provision of communications services at rates more closely related to costs; better management of communications networks, and provision of management expertise by users and intermediaries to the carriers; avoidance of wasted communications capacity; and creation of additional incentives for research and development of ancillary devices for use with transmission lines (Ibid., 265).

These expectations once more revealed the Commission's intervention on the side of corporate users. Cost-based pricing, on its face simply an effort to thwart monopolistic cross-subsidies, also suggests mounting concern to ensure that the telecommunications rate structure allowed major users to pay *only* for services rendered. In the very attempt to ascertain and control costs by seeking out such cross-subsidies, the FCC placed the entire rate structure of the telephone company at risk. This is not to enter the vexed debate over whether that rate structure was equitable; it is, rather, only to assert any forceful move to ensure cost-based pricing for one kind of service necessarily provokes a more general reevaluation of rate structure by the underlying carrier. The consequences of such scrutiny may of course impinge on other categories of service.

Growing demand to offer resale service, in fact, was directly an outgrowth of the Telpak rate disparity which permitted large customers to acquire private line circuits at bulk discount rates and gave them incentive to resell such circuits, with or without adding value, to customers who could not afford to take advantage of Telpak themselves. The Telpak discount, as Criner puts it (1977: 320), also "led to requests by user groups . . . for permission to secure these services on behalf of their members."* Resellers and sharers therefore would utilize private lines available to them at cut-rate prices. Such a trend was abhorrent to Bell not only because it undercut AT&T prices and forced new competition into its markets, but also because it played havoc with the whole strategy behind Telpak. Recall that Telpak had been instituted only as a competitive response to the threat of private networks posed by the result of the Above 890 proceeding; the Telpak strategy was reliant upon stringent restraints preventing users or competing carriers from,

---

*I should like to acknowledge Kim Price-Wen for this reference.

in effect, buying cheap and selling cheap. Large users therefore placed AT&T in a bind. Should the Telpak rate disparity remain, AT&T would effectively be sustaining its own competition. Should Telpak be eliminated, however, Bell would lose a vital weapon in its arsenal, use of which had helped to deter users from investing in private networks. The resale and sharing decision thus directly affected the precarious balance between public and private network development. (In its wake, AT&T tried the second alternative [in implicit recognition that a new competitive strategy was needed] and announced cancellation of Telpak discounts. Large users then responded by successfully making use of the courts to sustain Telpak for fully four years, until, in 1981, it was officially terminated [Brock, 1981: 209]. In mid-1981, 135 users of Telpak, including airlines, television networks, news organizations and the Federal Government, lost the average 28 percent discount ["AT&T Rate Rise On Four Services Set For Thursday," 1981: 7]. Further rate hikes for private line services were then authorized by the Commission, provoking a shrill outcry [and attendant legal action] from large users.)

This, however, by no means exhausts the implications of resale and sharing. To the degree that it nurtured new expertise in network management by resellers and nonprofit intermediaries like ARINC, the decision again concealed a hidden component. As Melody confirmed almost a decade ago (1973: 1265) "user telecommunications network planning, management and marketing functions are beginning to receive detailed examination of a type never experienced in the old monopoly environment." Beyond initial network design, new computerized techniques must be devised to diagnose and correct network operational problems of growing complexity. Performance standards must be developed and maintained. Configuration issues must be faced, particularly as networks are constantly changed to meet evolving user needs. Changes must, if possible, be anticipated without placing extra strain on the network, as organizational vulnerability to outrages increases together with functional integration of telematics into a widening field of corporate operations.

Network management services find extensive use in management efforts to create new forms of cost control. Reports generated in the system currently used by Atlantic Richfield (ARCO), for instance, permit this firm—operator of a large private network—to distribute telephone costs among different operating departments on the basis of actual usage. An Arco manager explains that the system can be used by department heads "to determine where opportunities for savings exist as well as to spot abuse situations" ("Information Service Tailors Cost Control System to Arco Unit's Individual Needs," 1980: 28). The system accordingly lists each call made by extension, including the number called, time and date, duration, class of service utilized, and the charge. Systematic violators of least-cost routing strictures admonishing employees to forego normal message telephone long distance circuits, for example, in favor of cheaper leased private line links, presumably are prime targets for lessons in telephone economics. Trade journals abound with examples of network management systems promising an infinite variety of like efficiencies across every industrial and service sector.

Crucially, the FCC hoped also to lay a foundation for further growth of private user networks. Underlying carriers would "benefit from the management expertise provided by resellers and nonprofit intermediaries to the using public" (60 FCC 2d 300):

> Whereas the underlying carrier is concerned with meeting the diverse needs of large segments of the using public, the reseller or non-profit intermediary will be concerned with the needs of a few user groups and should be in closer touch with those user groups than the underlying carrier. Consequently, the intermediary entity will develop better demand information than the underlying carrier, and this demand information will be conveyed to the underlying carrier when the intermediary orders communications services and facilities.

Not only would resellers wish "to employ the latest technological developments in order to make the most efficient use of the carriers' transmission capacity"—and thereby supply the carriers with "a new incentive to introduce new transmission technologies as soon as they develop, knowing that otherwise they may lose business as resellers" (60 FCC 2d 302). Most important, resale and sharing—because they vastly increased the legal ability of users and suppliers to identify and market specialized services making use both of computers and telecommunications, or what Melody (1973: 1266) has called "demand networks"—lay at the center of a new industry (60 FCC 2d 301):

> In short, we see a potential merging, in whole or in part, of the resale, specialized carrier and data processing industries into a highly competitive "information handling" industry, employing a combination of wholly-owned facilities, customer-provided facilities, and services and facilities provided by the underlying carriers.

Resale and sharing therefore catapulted users and suppliers both quickly toward an ultimate goal: fully merged, highly flexible, specialized computer communications. Before considering some aspects of an emerging industry structure in this volatile but vital area, we must dwell briefly on the ramifying competitive interests affected by this landmark series of decisions—to suggest that the FCC has today become involved in the most basic processes of corporate strategy and competitive position.

For, having once subjected private line service to resale and sharing, the Commission moved next to permit analogous use to be made of public switched services—including WATS as well as ordinary long distance or MTS. In October of 1980, resale and sharing restrictions on WATS and MTS were lifted because, the FCC asserted, they interfered with the right of users to employ these services in ways that were at once privately beneficial without being publicly detrimental—Hush-A-Phone again ressurrected. Benefits projected included new competition by firms offering sophisticated telecommunications management services in an expanded WATS/MTS environment, and new service options (Block,

"FCC Ends Ban On MTS, WATS Resale, Shared Use," 1980: 12). Users of these offerings of the underlying carriers were now free to expand innovative blendings of telecommunications and computers—subject, as resellers, to regulation—in any number of new ways.

AT&T had served notice on the FCC of its likely reaction before the decision to extend resale and sharing was taken. The telephone company filed for a rate increase in its WATS tariff that would transform the service to a more usage sensitive basis. The heaviest users of WATS, that is—the group standing to benefit from resale and sharing—would, under AT&T's new tariff, have to pay for their privilege. Employing logic which may have seemed ironclad to an agency preoccupied with cost-of-service pricing, AT&T announced that the new tariff was "designed to more accurately reflect the costs of providing each type of WATS service . . . to all customers" (Block, "AT&T files WATS revision, eyes private line boost," 1980: 12). While WATS rates would increase for the heaviest users, reductions would follow for most others.

Sensing the trend, perhaps, the Ad Hoc Telecommunications Users Committee (an elite group of the largest user firms including American Express, Control Data, Dupont, Ford, GE, Penney, K-Mart, Monsanto, Olin, Reynolds & Reynolds, Sears, U.S. Steel, Visa and Westinghouse) worried even in 1979 that AT&T might choose to eliminate or modify WATS after the fashion of its Telpak offering after the previous FCC resale decision (Comments, "In The Matter Of Regulatory Policies Concerning Resale And Shared Use Of Common Carrier Domestic Public Switched Network Services," FCC Common Carrier Docket No. 80-54, 17 September 1979). The Aerospace Industries Association (AIA) filed (Ibid. 9 May 1980: 1–2) to assert that "AIA's concern is particularly with respect to possible modifications in AT&T's WATS": both Outward WATS and Inward WATS (800 Service), declared AIA, "have become structural elements around which the communications systems of the aerospace companies have been designed." Even the *possibility* that AT&T might choose such a course impeded user planning and control. The Central Committee on Telecommunications of the American Petroleum Institute pointed out that the ironic consequence of more user intervention was in an immediate sense less predictability (Ibid. 9 May 1980: 3):

> users are presently facing a situation where telecommunications planning is impossible. This situation exists because all of the common carrier services provided by AT&T are under investigation and future services and rates cannot be determined at this time. This places the users in an extremely vulnerable position since without proper planning, millions of dollars can be uselessly expended.

The projected AT&T WATS tariff revision thus became the subject of a separate proceeding, wherein users again showed up in force to demand a hearing. The petition by Avis Rent A Car System, Inc., to suspend the proposed tariff, furnishes a glimpse into some of the questions and stakes involved. Avis had recently consolidated its reservation system in Tulsa, Oklahoma, "in order to provide improved customer service" through one nationwide toll-free Inward WATS number for car reservation service ("In the Matter of American Telephone And Telegraph Company Revisions To Tariff FCC No. 259 (WATS)," FCC Common Carrier Docket No. 80-765, 15 October 1980: 1). Through acquisition of multi-million dollar call-distribution equipment to attach to its WATS lines, Avis expected to optimize its use of 800 Service. In other words, the centralization of the entire Avis reservation system was dependent upon WATS. If the proposed tariff took effect, however, "this investment will be for naught as the proposed tariff does not reward efficient use of the WATS/800 Service." Specifically, if implemented the new tariff would raise Avis service and equipment expenses for these facilities from $255,000 per month to $421,000 per month—a 65 percent increase.

Paradoxically, the ubiquity of AT&T's network meant that the largest users—the group most demanding competitive alternatives—were most dependent upon AT&T, especially in regard to consumer-oriented services. As Citicorp suggested (Ibid., 20 October 1980: 1), the proposed tariff "may be more a response to market demand factors, including the presence of competitive alternatives for many smaller users and the absence of such alternatives for many larger users, than an attempt to formulate cost-based rates." If Avis utilized WATS for its nationwide customer reservation system, Citicorp employed it in part to provide authorization and customer/merchant services to several million VISA, Master Charge and other credit cardholders and merchants. Citicorp asserted (Ibid.: 1–2) that the proposed tariff restructuring could result in increased charges in excess of $2 million annually and, in addition, "may render unusable recently acquired computer systems designed to optimize telephone traffic based on the *current* rate structure." Projecting, Citicorp saw total telecommunications cost increases of 50–100 percent at some locations and complained that it "cannot absorb the staggering telecommunications cost increases associated with this AT&T request without passing a major portion of such costs on to cardholders and merchants" (Ibid.: 2).

Perhaps the most eloquent general appeal was made by the Committee of Corporate Telephone Users (CCTU), formed in 1973 and currently numbering some 35 users including Avis, Time Inc., Greyhound and Citicorp, in its petition to reject or suspend the WATS revisions (Ibid., 20 October 1980: 8). "Since the introduction of WATS," the CCTU charged (outbound WATS was introduced in 1961; 800 Service in 1967), "companies of all sizes have come to depend on it." Indeed,

> In reliance upon this bulk rate offering, firms in a multitude of industries located throughout the nation have made enormous capital investments in telecommunications equipment; they have fundamentally altered their business operations and plant locations; and, they have offered new and more efficient services to the public. Indeed, many companies owe their existence entirely to WATS. . . .

Reliance upon WATS in corporate planning was in keeping with the larger incorporation of telecommunications as an essential basis for business operations; "In the field of banking," the CCTU filing underlined (Ibid.: 11), "communications has, in effect, replaced the gold standard."

While proliferating lawsuits attest the preeminence of communications in various business offerings and services, the implications of the resale and sharing decisions themselves are today just beginning to be developed. Two prominent Western International hotels will be reselling WATS and long distance services to guests, routing calls over least-cost Bell System circuits while charging customers at a profitable rate ("Two hotels to resell phone service for profit under new FCC regulation," *Telephony* 4 May 1981: 105). Many other users, predicts one analyst, "will build their own phone companies"—for resale and/or sharing ("At the Office with Howard Anderson: The View From The Sidelines," 1981: 3). By virtue of the decision to permit resale and sharing of WATS and MTS "companies selling such discounted long distance services are no longer limited to the cities in which they have leased lines and installed equipment"; for they now are able to provide a variety of new services nationwide ("The resale business in phone lines," 1981: 68). More than thirty resale companies have started up; thirty more have applied to the FCC for permission to do so. Sharing arrangements, largely unregulated, are not as easily appraised. Faced with large increases in telecommunications costs, however, some businesses are seemingly looking to share facilities. Bank of America and Firestone Tire & Rubber, for example, may both be moving in this direction ("ICA President Lloyd Isaacs Discusses Bank and Association Goals," 1981: 54; "The resale business . . ." 1981: 68). A new trade association, finally, whose principal purpose will be to promote "the common business interests of those engaged in the shared use, network management and resale industries," has already been formed ("Form Group for Persons Involved in Resale Transmission Business," 1981: 15).

Because the series of resale decisions implied that value-added services henceforth might be contracted for and supplied as unified telematics packages—that is, that one entity might provide both telecommunications and data processing, merged as required by a given application, and subject to regulation—it must ultimately be judged in terms of the growth of merged value-added services. Let us now then briefly consider the structure of this highly dynamic applications-oriented "information-handling" industry, that the FCC has done so much to cultivate.

Tymshare, a leading computer services firm, provides solutions to what the company calls in its 10-K Report to the SEC "diverse information retrieval, information management and computational problems" (Tymshare Corporation 1979: 1). Concentrating principally on tax accounting, financial, health care, manufacturing industries, utilities and government, Tymshare processes credit cards for Mastercharge and Visa, performs nationwide automated credit card verification for airlines, car rentals, hotels and restaurants, provides financial and patient tracking services for hospitals, and is involved in inventory, shop floor control, order entry and billing, and financial accounting for "a large number of major domestic corporations in the petroleum, petrochemical, electronics and aerospace industries" (Ibid.: 2; "Top 10 At A Glance," 1980: 11). Tymshare owns and operates Tymnet, a data communications network reliant upon existing standard telephone lines whose revenues are growing at 50 percent a year (compared to 30 percent for Tymshare as a whole) (Ibid.).

Tymshare, however, is but one of a large and diverse group of information service providers catering to virtually all business and industry sectors. Quickly thickening societal dependence on such companies is glimpsed in testimony of Thomas O'Rourke, a member of the board of directors of the Association of Data Processing Service Organizations (ADAPSO), and president and chairman of Tymshare ("Statement of Thomas J. O'Rourke," *Status Of Competition And Deregulation In The Telecommunications Industry*, Hearings Before The Subcommittee On Telecommunications, Consumer Protection, And Finance, Of The Committee On Energy And Commerce, House of Representatives, 97th Congress 1st Session, 20, 27, 28 May 1981. Serial No. 97-29, USGPO: 314):

> Today the list of applications is endless. Our customers utilize our services to operate steel mills and petroleum refineries, to control production, and to perform any or all of their accounting functions. Many Fortune Five Hundred industrial customers, in spite of the fact that they have extensive inhouse computers, use our interactive services to perform market and statistical analyses and financial modeling on line from terminals. Engineers use our services to design bridges, buildings, and atomic reactors. Banks have teller terminals connected to our computer centers to provide all types of financial services. The use of our services by customers to maintain data bases for inquiry is limitless. Educators utilize our services for computer based education systems. Federal, state and local governments perform all types of statistical, accounting and engineering applications by means of our services. In summary, almost any business, educational institution or government entity, large and small, is in some way being provided information services by our industry.

Fully 11,000 companies are actively combating for pieces of an information services market estimated to be worth $12.5 billion (Ibid.: 315). Fewer than 50 industry participants have annual revenues in excess of $25 million (Ibid.: 316). Many comparatively small firms develop and market diverse specialized or customized applications to meet a myriad expanding customer needs (Blumenthal, "Processing Services Now A Whole New Ball Game," 1980: 1, 10).

The current explosion in information services grows directly out of the computerization of U.S. business as a whole. The renowned success of the computer industry in dropping prices—the cost of computing, *Business Week* estimates, has declined 20 percent or more each year for two decades ("Missing Computer Software," 1980: 46)—has resulted in proliferation of computer hardware. Paradoxically, the very diversity of contexts in which hardware is situated engenders a powerfully escalating demand for more, better, specialized software. New users demand new programming to do new things with computers. The increasing range of information service products, declares an industry spokesman, "is necessary to reflect the needs of an ever-growing number of consumers, since no one package or service could be developed to fit all of a customer's requirements" (Thomas J. O'Rourke, in *Hearings On The Status Of Competition And Deregulation* . . . . 1981: 316) It is not, as has sometimes been claimed ("Missing Computer Software," 1980: 49), that there has developed a sudden "shortage" of programmers—but that this shortage has itself grown from the ever-widening use of computer hardware. Today therefore there is a high premium placed on those who can furnish customized services to solve an enlarging series of applications problems.

The sources of expertise in this field are three: computer companies; specialized services firms, like Tymshare, set up to furnish "solutions" to problems of particular applications or industries; and users themselves, who may be positioned to translate their hard-won expertise into an independent source of profit. Who knows the specific problems of banking better than the banker? Once in-house computerization has attempted to solve some of these, it is then but a short step to providing answers to other banks. Finally, proliferating service offerings create yet more demand for data communications, without which centralized software, database and maintenance and diagnostic resources could not be easily accessible. In 1979, 60 percent of processing services revenue derived from use of systems incorporating data communications; this figure is expected to grow to 75 percent by 1984 (Blumenthal "Processing Services Expected to Grow 17 percent/Year," 1980: 12).

Let us examine one sector of the information services market, processing services (which allow remote access to software and databases), somewhat more closely for, though initially it may seem unrelated, it benefits directly from FCC resale and sharing decisions. In 1979, of $7.6 billion revenues accumulated by some 4,000 processing services companies, the top twenty firms received about one-third ($2.53 billion) (Table 5). Of this elite group, twelve were affiliates of Forbes 500 companies in electronics, aerospace, computers and business equipment, banking and finance—precisely the industry sectors represented in proceedings we have scrutinized at the FCC. Providers of processing services—critically reliant upon telecommunications to deliver service offerings—may enter this value-added line of business from telecommunications, computers, or wholly other fields.

This may reflect either a spin-off of originally in-house data processing resources or acquisition of services firms as part of broader corporate strategies, or a combination of both. United Information Systems, McDonnell Automation and Boeing Computer Services (and, prominent though not on this list, Sun Information Systems and Citicorp's Citishare subsidiary) are instances of the first tend-

TABLE 5
Top 20 Processing Services Firms

Company	1979 Processing Service Revenues ($ millions)	Market Share (%)
Control Data Corp	420	7
Automatic Data Processing	409	7
General Electric	290	5
Tymshare	160	5
United Information	138	3
McDonnell Douglas Automation	135	2
Computer Sciences Corp	129	2
NCR	93	2
Bradford National	89	2
National CSS (Dun & Bradstreet)	77	1
Bank of America	70	1
Comshare	67	1
Mead Data	65	1
Boeing	62	1
TRW	55	1
University Computing (Wyly)	60	1
Xerox	60	1
National Data	54	1
Trans Union System	50	1
First Data Resources (American Express)	50	1

Source: Marcia Blumenthal, "Processing Services Now A Whole New Ball Game," Copyright 1980 by CW Communications Inc. Framingham, MA 01701-reprinted from *Computerworld*.

ency. They exemplify a lengthy telematics tradition; Southern Pacific Communications, TRT and Radio Liberia are examples of telecommunications companies begun first as in-house concerns by Southern Pacific Railraod, United Fruit and Firestone Tire and Rubber, respectively. On the other hand, acquisition of First Data Resources, National CSS, System Development and Interactive Data, by American Express, Dun & Bradstreet, Burroughs, and Chase Manhattan, respectively, are illustrative of the second tendency. Both trends are pronounced today; information services has become an intensely volatile industry. Data processing services industry mergers and acquisitions in 1979 accounted for 10% of all U.S. merger activity in that year (107 transactions valued at $671 million). According to one study in a five year period from 1975, the number of acquisitions of computer services companies increased by 168 percent—making it the leading acquisition-prone industry (Barna, 1980: 50–52). Figures for 1981 indicated that 118 separate merger or acquisition transactions, valued at $766 million, occurred during one year ("Index: 118 Mergers, Acquisitions Occurred in '81 Services Industry," 1982: 79). Together, software and services industry sectors are growing at a phenomenal 30 percent a year (Blumenthal, 1981: 73-5).

The largest processing services firm in 1979—and number three on the Datamation 100—Control Data Corporation embraces a very substantial services component. Currently Control Data operates Arbitron, Ticketron and Cybernet (entertainment ticketing services), timesharing services with over 2,000 applications, banking, credit union, brokerage, electronic funds transfer and delinquent accounts collection data processing services, database services (Evans Economic Data Base) and educational services (PLATO) ("Top 10 At A Glance," 1980: 11; "Datamation 100," 1981: 105). Another prominent computer firm, Burroughs, purchased System Development Corporation in January 1981 for $98 million to provide the parent "with additional financial and marketing clout in the software and services market." Deriving 35 percent of total data processing revenues from software and services (about $867 million in 1980), Burroughs expects its new subsidiary "to help increase its government business, which only accounts for 5% of Burroughs' revenues" ("Datamation 100," 1981: 154, 108). In the wake of the antitrust suit against IBM—dropped by the Justice Department in January 1982—the telematics giant has announced its decision to reenter the computer services market via a remote computing facility called the IBM Information Network (Chace and White 1982: 2).

Entry into the services market is also open to telecommunications companies. GTE Telenet, a packet switched network now within local dialing distance of most businesses in the U.S., is a case in point. Its Financial Services Division provides services to the investment community, using a 100,000 mile global quotation network to do so; this service "enables brokerage firm personnel to have all available public and private brokerage information at their disposal through desk-top display terminals, and permits instant communication within brokerage firms" (General Telephone and Electronics Corporation 1979: 6). United Information Systems—a spin-off of United Telecommunications, one of the largest independent telephone companies—provides management information systems and services, telephone and engineering applications, and government services ("Top 10 At A Glance," 1981: 11). Then of course, there is AT&T, whose long-delayed Advanced Communications Service—recently announced—is intended to be an encompassing data communications network service, permitting users to interconnect otherwise incompatible terminals to receive a wide range of projected Bell services (Schultz, "ACS: Will It Be Baby Bell's Problem Child?" 1981: 5). The unsurpassed importance AT&T attaches to value-added services is shown by its shift from the telephone business to what current ads call "the knowledge business," and is perhaps most dramatically evidenced in its recently proposed antitrust settlement with the Justice Department—of which more in a moment.

Despite massive involvement by data processing and telecommunications companies in provision of services, this mushrooming sector is in no way confined to them. Specialized applications of software and services rely upon ready access to substantive information. Virtually any organization or entity with access to market, commercial, corporate, financial, consumer, employment, investment, tax, demographic, sales, advertising, political and all other sorts of data, thus, is a candidate for entry into the services field. Informational advantages are quickly becoming a key factor in restructuring markets; coupled to telematics technology, they confer an independent source of market power on those who are in a position to exploit them; and, although the final results are not yet in, it is apparent that this newfound source of market power is finding extensive use in furthering economic concentration within U.S. industry. Within industries, firms positioned to harness special information advantages are changing the shape and structure of the markets in which they operate. More broadly, certain unexpected industries and industry sectors are mobilizing to invade selected services markets. Sears' purchases of Dean Witter Reynolds, the nation's fifth largest brokerage house, and Coldwell, Banker & Company, the country's largest independent real estate broker, would seem to demonstrate the parent firm's intent of drawing upon its unparalleled resources in retailing to become the "largest consumer-oriented financial services entity" (Cole, 1981: A1). The project reveals its informational aspect when it is recalled that Sears, in addition to owning the giant Allstate Insurance Company, has extended some 24 million Sears credit cards to consumers (Ibid.). Presumably data generated through records of credit card use will facilitate untold new services in conjunction with Sears' recent acquisitions.

Sears will probably have to confront other rising stars in financial services if it is to succeed in its plans. Notable in this regard will be the expansion of banks and other financial service firms. In this field there has been an unprecedented wave of huge mergers of late—Salomon Brothers with Phibro, Prudential Insurance with Bache, American Express with Shearson Loeb Rhoades. Probably furthest advanced is Citicorp. Through its Citishare subsidiary it acquired at least seven computer services companies in 1980–1981, forming numerous other subsidiaries to capitalize on what was at first an in-house computer processing resource. Having been what *Datamation* calls "the single largest driving force behind the move to distributed processing in the 1970s, Citibank"—Citicorp's banking subsidiary—"clearly wants to profit by selling this expertise in the 1980s" (Emmett, 1981: 47). The "position of privilege" held by the largest U.S. banks derives, in this view, not only from previous investments in telematics technology, but from investment relations with other companies and, in particular, from being conventionally "privy to all kinds of sensitive information from their customers and the government alike" (Ibid.: 48). Citicorp, that is, has an informational advantage that is part and parcel of its ensconced position as one of the world's leading banks. The same is true in other sectors. Thus, according to this trade journal, information services are being invaded by giant firms, each attempting *both* to centralize control in its own field but also to "hunt outside its own traditional turf" (Ibid.).

No matter who provides them, however, information services are characterized by what O'Rourke of ADAPSO labels "a major dependence on telecommunications" (*Hearings On The Status Of Competition And Deregulation* . . . 1981: 316) and resale of private line service, WATS and MTS by information service firms is crucial to further development of this fused, applications-oriented industry. Yet resale carriers, recall, *were still subject to regulatory oversight by the FCC*. Would telematics develop under continued regulation, as the latter was extended to integrated information services—or would existing regulatory strictures be relaxed to permit integrated telematics services to emerge unencumbered, as private enterprises pure and simple? Closely related was the status of AT&T: on what terms would the regulated core of the nation's telecommunications network be allowed entry into the blossoming resale environment? With these fundamental issues we approach the heart of the current debate.

Competitive Common Carrier and Computer Two

The conceptual intent of the Competitive Common Carrier proceeding was nicely described by the Central Committee On Telecommunications of the American Petroleum Institute; prospective FCC action was based on "two fundamental concepts" (FCC, "In the Matter of Policy and Rules Concerning Rates for Competitive Common Carrier Services . . . ," Docket No. 79-252, 1 May 1981: 3):

> First, the costs of regulation should not outweigh the benefits of regulation that are intended to inure to the public; to the extent the costs of regulation exceed the benefits, the regulations or the application thereof should be modified. Second, communication carriers that lack market power, carriers that, in the absence of regulation, could not maintain or control the price of a service(s), should not be regulated in the same manner as those carriers possessing market power.

The proceeding was undertaken in an attempt to relieve non-dominant carriers—that is, those unable to engage in unlawful cross-subsidy of services—of certain specific regulatory requirements. Specifically, the Commission thought to shorten the notice period for tariffs filed by non-dominant carriers and to relieve the latter of the burden of certain tariff filing requirements. Should it result in a favorable verdict, however, this seemingly innocuous inquiry would decisively support both merged or value-added services and already extensive private networks previously ratified by decisions to permit resale and sharing.

For, in 1981, the Competitive Common Carrier proceeding was broadened substantially (84 FCC 2d 445). Through a Further Notice Of Proposed Rulemaking the Commission decided to consider whether it might forbear *entirely* from regulation of non-dominant private carriers—that is, those carriers lacking substantial capability for exercise of undue market power—including resale carriers. Dow Jones, for example, which made use of an extensive mix of owned and leased communications facilities, had made no attempt to share them—this being thought unprofitable—or to resell any excess capacity. Resale had been rejected, Dow Jones appealed (FCC "In the Matter of Policy and Rules Concerning Rates for Competitive Common Carrier Services and Facilities Authorization Therefor," 27 April 1981), because under current rules it would elevate the company to common carrier status before the Commission and hence subject it to regulation. However, Dow Jones would surely enter the market as a resale carrier were the Commission only to exempt such carriers from regulation. Such a move would cap the long-term effort to privatize domestic telematics. Should the Competitive Common Carrier proceeding be carried to its seemingly natural conclusion, in June 1982, the very concept of common carriage may well be placed at risk.

This of course is what prompted the American Petroleum Institute, for instance, to object to total deregulation of non-dominant carriers. So, too, did the Association of American Railroads. Their opposition stemmed from a longtime reliance on what today are called Private Radio Services, which make use of the electromagnetic spectrum. In 1981 the railroad industry employed more than 41,000 route miles of private microwave; "Over the years, these private communications systems have been integrated into the day-to-day railroad operations and have become indispensable for the safe and efficient conduct of railroad business" (Comments of the Association of American Railroads Ibid. 1 May 1981: 2). "Without those systems, railroads simply cannot function" (Ibid.). Squatters' rights dictated that private microwave networks—here and in other industries—should not be jeopardized. Should the Commission through deregulation of non-dominant common carriers erode or even eliminate its long-standing policy distinction between common carriage and private radio, however, the frequencies currently reserved for private radio—already chronically scarce—might no longer be granted to this exclusive use. Thus the largest users of private radio worried over the implications for their continued rights over a portion of the spectrum, should a class of "non-dominant" deregulated carriers be mandated.

What—finally—about the *dominant* carrier? In the Second Computer Inquiry, the other profoundly vital aspect of any transition to private telematics—the role of American Telephone & Telegraph—was the major topic of debate (FCC, "In The Matter Of Amendment Of Section 64.702 Of The Commission's Rules And Regulations (Second Computer Inquiry)," *Final Decision* Released 2 May 1980).

We have seen that the First Computer Inquiry, resolved early in the 1970s, created a fundamental division between unregulated data processing and regulated communications. It left what Brock (1981: 270) calls "an arbitrary line" which prevented any fluid merger of computer and communications technologies. Decisions to authorize value-added networks coupled with the later resale and sharing decisions signified that any such dichotomy was now technically obsolete. Economic demand for integrated services forced the issue by making the distinction also pragmatically unenforceable. Something had to give. In 1976, thus, the Commission undertook a second inquiry into the proper dividing line between data processing and telecommunications (Ibid.). Over the course of the Second Computer Inquiry the decisions to permit resale and sharing and to expand the Competitive Common Carrier proceeding were taken, laying the groundwork for a new telematics policy framework. Simultaneously, the previous dichotomy was attacked from both sides. While, as we have noted, computer companies, users and value-added services firms merged data processing and telecommunications in innumerable concrete applications, the telecommunications companies (AT&T especially) produced more and more computerized equipment for incorporation into the nation's telephone system. If resale and sharing and Competitive Common Carrier proceedings helped set down the basis for a new telematics policy founded on the demands of U.S. business as a whole, the Second Computer Inquiry dealt specifically within this evolving framework with the crucial question of the role to be assumed by AT&T.

Comparatively early in the proceeding the Ad Hoc telecommunications Users Committee—a group of the very largest users embracing Bethlehem Steel, Dupont, Ford, Monsanto, Olin, Republic Steel, Union Carbide, Shell Oil, Westinghouse, Penney, American Express, Chrysler, Exxon, Sears and U.S. Steel—filed an extraordinary comment. This filing broadly but explicitly underscored that the very freedom of large corporate users to expand and develop telematics was in turn contingent on the carriers', and most especially on AT&T's, ability and willingness to upgrade the public network. As the owner and operator of the nation's core telecommunications network AT&T should be permitted, users insisted, to modernize its facilities—thus preparing the way for users' employment of data communications on an unfettered national scale. The Ad Hoc Committee urged the Commission (FCC, "In The Matter Of Amendment Of Section 64.702 Of The Commission's Rules And Regulations (Second Computer Inquiry)" 16 May 1977: 3–4), "to assure the ongoing freedom of communications common carriers to utilize new kinds of equipment and new processes in providing communications services." The group argued that communications as defined by the 1934 Act were dynamic, and very precisely intended to assure that, in preference to the public as a whole, "the communications-using public" specifically, might enjoy their benefits. Carriers, that is, should have opportunity "to provide, with carrier-supplied equipment, all types of communications services which may be found useful'' (Ibid.). No limit, the users decreed, employing the already outdated distinction between data processing and communications to make their point, should be placed "on the freedom to innovate in providing *communications* services"—even were the techniques to be employed by innovating carriers borrowed from unregulated industries. In a powerful phrase the users attacked "technological segregation," which, they asserted, "if practiced now, could deny to the customers of communications common carriers the benefits of the use by the common carriers of any equipment or process within the scope of a broad definition of data processing functions" (Ibid.: 5). The Committee's concern

> is that the Commission assure the freedom of carriers to provide a complete communications service involving long-term evolutionary improvement in communications processing equipment, and that the prohibition against car-

rier intrusion into the data processing business does not obtrude on that freedom (Ibid.: 5–6).

And the Committee singled out AT&T for especial attention. It was "extremely important to users," the group declared, "that the considerable talents of AT&T and its subsidiary companies be devoted to the solution of data transmission problems" (Ibid.: 10). Then, the clincher; despite the progressive merger of data processing and telecommunications, it was *vital to users "that the Bell System be permitted to provide unregulated data processing services* when, because of blurred distinctions, this becomes necessary to fulfillment of its future communications responsibilities" (Ibid. My emphasis). New legislation should be enacted "to make certain that an AT&T subsidiary may lawfully offer non-regulated data processing services, when such services are offered pursuant to Federal Communications Commission exemptions from regulation" (Ibid.). Integral telematics services should be made available without delay on a deregulated, private basis. Because provision of such services by AT&T was essential to corporate users having need of its peerless national network, even AT&T should enter the market through a deregulated subsidiary or affiliate.

> The Commission should act to protect the user public, and should favor the broad user interests over those of any particular supplier or supplier group. The user public's interests demand free availability of terminal and incidental communications equipment whether supplied by carriers or others. Before *Carterfone*, terminal and incidental communications equipment for connection to the public telephone network could be secured from common carriers only. Since *Carterfone*, alternative sources have developed. This arrangement has been most helpful to users. The Commission must guard against adopting technological segregation. Its efforts to prevent carriers from providing non-communications services in violation of Commission policies can be enforced without denying carriers and their communications customers the benefits of evolving technological improvements in processes. To fail to guard against technological segregation might deny users ready future access to important improved services made possible by new technology in terminal and incidental communications equipment offerings. Excluding common carriers from supplying services utilizing innovative communications terminals would surely have that result, because such machines are a wave of the future. Carriers as well as independent suppliers should be permitted to develop them. The danger of technological segregation exists, unless it is made clear, as in the proposed rule changes, that common carriers may continue to furnish evolutionarily improved communications terminal and incidental equipment, as appropriate to respond to all communications needs of users, over the long term. (Ibid., 16 May 1977: 21–22)

The underlying issue of the Inquiry, therefore—despite the labyrinthine complexity of many associated questions—was rather simple. How could the national telecommunications network dominated by AT&T be upgraded to enable extensive and efficient data transmission without causing telematics to come under more regulatory supervision? However much corporate users may have worried over the ramifications, they simply needed AT&T too much as an underlying carrier to oppose provision of unregulated services by the Bell System. Because it was critical to users to keep telematics from becoming "ensnarled in needless regulation"—the term is then FCC Chairman Charles Ferris'—AT&T had to be given rein (Ibid., *Final Decision*, Separate Statement Of Charles D. Ferris, 7 April 1980: 2).

After an unsatisfactory tentative decision by the FCC in July 1979, offered another unworkable version of the dichotomy between data processing and communications, in the process provoking widespread complaint owing to its potential expansion of regulatory controls into the previously unregulated data processing area, the break came. In its first "final" decision in May 1980, the Commission set out to deregulate all terminal equipment including ordinary telephones as of March 1982. After that, customer premise equipment might be furnished by the two "dominant" telecommunications firms only through a separate unregulated subsidiary, and not as a part of their basic telephone service. The decision thus rests on a new distinction, between "basic" services that simply move information in real time, and "enhanced" services that combine the former with computer applications to provide additional, different or restructured information. Computer-enhanced transmission services, according to this decision, could also be provided by the two dominant carriers by means of separate subsidiaries that purchased basic transmission capacity on a tariff basis as resale carriers. Non-dominant smaller carriers were granted freedom to furnish enhanced services and terminal equipment without making use of such an arms-length subsidiary. The key result of Computer II has been to plunge AT&T into integrated telematics via new, unregulated subsidiaries which will compete with other private resellers.

Much litigation has ensued. At the Federal Communications Commission, GTE has been exempted from dominant carrier status and may therefore offer enhanced services without setting up a separate subsidiary. The FCC has held that its jurisdiction does in fact extend to common carriers providing enhanced services insofar as protection of customers in respect of other, still regulated carrier offerings is concerned. Concrete markers with which to divide regulated from unregulated services remain in dispute, as is the degree of separation between regulated and unregulated segments of the Bell System.

In Congress, meanwhile, the terms upon which the Bell System shall be permitted to furnish integrated telematics equipment and services have likewise become pivotal. In very large part, efforts to rewrite the 1934 Communications Act center on this fundamental issue. As passed by the Senate in October 1981, S. 898 substantially frees the Bell System to enter hybrid telematics markets, although a number of conditions were placed on AT&T's activities in such unregulated fields as a result of concerted lobbying by user groups and equipment manufacturers and independent competitors. The House bill—H.R. 5158—subjects AT&T to more rigorous regulatory oversight, and links deregulation of a given service market to the actual degree of competition prevailing within it, while still moving swiftly toward a deregulated telematics environment.

The Second Computer Inquiry, though still subject to substantial judicial interpretation, gave AT&T freedom to embark on a network-wide attempt to move into unregulated, merged computer communications. This the company lost no time in doing. Corporate reorganization to engender an aggressive new subsidiary, "Baby Bell," accompanied resolution of the Second Computer Inquiry—in spite of the fact that this FCC decision proved to be far from final. New service offerings, most notably a value-added data communications network called the Advanced Communications Service, are being rapidly readied to compete in deregulated telematics markets. Taking every advantage of the result of the Second Computer Inquiry—a result to which business users had contributed—AT&T moved aggressively to prepare for the coming privatized environment. By Summer 1981 a report on AT&T's ongoing restructuring concluded that *"Bell has taken the initiative away from the regulators and competitors and now in large measure controls its own destiny"* ("Yankee Group Sees Changes in AT&T Goals," 1981: 18, original emphasis).

Business users quickly recognized that, absent a major effort to force restrictions on AT&T by legislative means, the regulatory mandate supplied to the company by Computer Two might place them once more at a comparative disadvantage. Asked if he feared that "a deregulated AT&T would have . . . power to hurt you?" Dan Grove of the Tele-Communications Association replied ("Mike Woody and Dan Grove Tell of TCA Plans and Progress" 1981: 54):

> I guess it's my feeling that they would have within their power to pursue various marketing policies that they feel would be in their best interest. You have to assume that anyone is going to go out and take advantage of the strengths that they have and if they can do anything to discourage competition, you would expect them to do that, as would anyone. And the Bell system management has been quoted as saying that profit is going to be the goal of that deregulated subsidiary. They are no longer out now just to make a good rate of return but they are going to be looking at that profit margin in the bottom line and whatever they can do to achieve that goal, I believe they will do . . . . So in that respect I'm concerned about what impact it might have on me as a service procurer from the Bell System. For example, in a circuit I may have between Phoenix and Seattle I may wind up paying a rather high price because there is no competition. On a circuit between Phoenix and Dallas there may be a much lower rate. They would have that ability. The other side of it is that if they are successful in at least hindering the development of competition, then I would feel that I would be inadvertently or indirectly affected by that.

Users once again therefore stepped into the breach. Specifically, in 1981 a broadbased coalition representing over 5,000 companies that use, provide or manufacture telecommunications and information services or equipment—telecommunications Competitive Alternatives for Users Services and Equipment, or Tele-Cause—was formed. Tele-Cause constitutes the most general and extensive organizational move to check the prerogatives of the Bell System ever attempted. Members expend about $20 billion annually on telecommunications services and equipment, and represent "virtually every segment of United States industry" ("Coalition of Communications User Groups Forms Tele-Cause," 1981: 78). The group also comprises the most thoroughgoing alliance between equipment manufacturers, service suppliers and business users. Members include ADAPSO, the principal trade association of the computer services industry; the ICA, the largest business user group; the Independent Data Communications Manufacturers Association, representing a small but rapidly growing group of equipment manufacturers; Control Data Corporation; International Telephone & Telegraph Corporation; General Telephone & Electronics Corporation; Satellite Business Systems—the consortium formed by IBM, Comsat and Aetna Life & Casualty; the National Retail Merchants Association, whose 4,000 members now operate 40,000 stores across the nation; and the Tele-Communications Association. The group's general aim is to ensure that where the "Bell System is to be allowed to enter unregulated markets, such as information or data processing," legislation "must ensure that real competition, not new monopoly, is the result. Deregulation must succeed, not precede, true competition . . ." (Ibid.). Numerous more specific goals in respect of any legislatively-mandated transition to an unregulated telematics environment were formulated; as the Fall of 1981 wore on, Tele-Cause and its diverse members brought pressure to bear on the House Telecommunications subcommittee to support their interests with a user-oriented bill.

When introduced into the House of Representatives, H.R. 5158—as mentioned above—did indeed subject AT&T to regulatory constraints and safeguards absent from its companion bill in the Senate. Conflicting provisions between the two bills guaranteed both delay and the prospect of heightened intervention by the user-manufacturer alliance. It was in this context of intense and still mounting pressure that AT&T sought to protect its bold move into unregulated telematics, and to preserve its increasing competitive momentum, by dramatically resolving yet another contest of profound importance to its corporate future: the antitrust case levied against the Bell System by the U.S. Department of Justice in 1974.

The Antitrust Case

The story of the recently concluded Government antitrust case against the Bell System leads directly back to an earlier suit brought in 1949 as part of a last-gasp effort by outgoing New Dealers to check monopoly power in the telephone equipment industry ("Truman Plants an Antitrust Time Bomb," 1948: 19; "The New

Attack On Bigness Takes Shape," 1949: 26). After several years of proceedings, a consent decree was obtained. AT&T was required to make its patents available to all applicants, thereby denying it a formidable—and often utilized—source of market power. AT&T was likewise foreclosed from participation in unregulated markets, which at the time comprised just a tiny fraction of the firm's business. Western Electric, AT&T's manufacturing arm, was also prohibited from producing equipment other than for the Bell System and the Government. Hailed by Government attorneys as a major victory, the decree of 24 January 1956 was in fact a triumph for AT&T. As a House subcommittee later conceded (in Uttal, 1981: 72), "The consent decree . . . stands revealed as devoid of merit and ineffective as an instrument to accomplish the purposes of the antitrust laws."

The victory unquestionably belonged to Bell. The major item of relief sought by the Justice Department—breakup of AT&T's tightly integrated corporate structure, including divestiture of Western Electric—fell by the wayside.

Continuing technical advances in microelectronics, coupled with mounting demand by users for merged telematics offerings in unregulated markets, ironically turned the 1956 consent decree into a crucial fetter on AT&T. Not the patent-licensing provision (which had been the Government's chief claim to victory in the case), but the prohibition on entry into unregulated markets, became its most vital constraint. In 1956, of course, it could not have been foreseen just how intimately intertwined computers and telecommunications would become; the integrated circuit had not yet even been invented. Twenty years later, however, the situation was radically changed. Unless Bell could somehow vitiate the terms of the decree, declared AT&T attorney George Saunders, in a bit of calculated overstatement, "we are a withering corporation waiting for its demise" ("Bell 'a Withering Corporation' Under Consent Decree, Attorney Says," 1982: 11).

Stemming not so much from trustbusting conviction but from the clamorous complaints of competing equipment suppliers, specialized common carriers, and business users, a second antitrust suit was levied against the Bell System on 20 November 1974. The complaint addressed "the recurrence in the last three decades of anticompetitive activity" by AT&T (Civil Action No. 74-1698, In The U.S. District Court For The District Of Columbia, United States Of America, *Plaintiff*, v. American Telephone And Telegraph Company; Western Electric Company, Inc.; and Bell Telephone Laboratories, Inc., *Defendants; Plaintiff's Third Statement Of Contentions And Proof*, 10 January 1980: I, 5). As other companies had responded to "substantial demand for additional new telecommunications services and equipment" in the years after World War Two, the Justice Department asserted, AT&T had relied on its position of "dominance in long-distance transmission, equipment manufacturing, and local franchise monopolies, and the leverage derived therefrom, to suppress this new competition and to maintain and enhance (its) market power" (Ibid.). In instance after instance of what it claimed comprised anticompetitive conduct, the raw materials for the Government's case were harvested from the experience of business users, independent equipment suppliers and specialized common carriers. The suit thus became a third critical site, with the FCC and the Congress, of the contest between users and the Bell system.

Relief sought by the Government centered on splitting AT&T into regulated local telephone companies and several unregulated suppliers of telecommunications equipment and long-distance services. Thus its chief goal was creation of a privatized or unregulated telematics environment protected from excesses of market power by dismemberment of AT&T's integrated vertical structure.

On 8 January 1982, the first half of this goal was provisionally attained. Linking the settlement to the outcome of the previous 1949 antitrust suit by way of a high-priced legal manuever, AT&T with one bold stroke accomplished what the FCC and Congress had labored so long in their different ways to achieve for business users: provision for merged telematics equipment and services *without* accompanying extension of regulation. Members of the FCC immediately lauded the settlement, as Chairman Mark Fowler voiced characteristic hope that "in the long term, the decree will make it easier to unregulate large segments of the telecommunications industry" ("Progress, Many Problems Cloud AT&T-DOJ Settlement," 1982: 15).

Settlement of the case occurred in the form of a "Modification Of Final Judgment" *on the 1956 consent decree*. Under the terms of the proposed agreement, the 1956 decree was vacated in its entirety, and replaced by a series of new provisions; the current (1974) antitrust action was at the same time summarily dismissed. Final approval of the 13 page document which outlines only the most general principles to be followed by parties to the agreement, must still be accorded by Federal Judge Harold Greene. A more detailed plan must also still be submitted by AT&T to the Justice Department, for the settlement to be concluded.

Even the very general outlines of the proposed settlement, however, are of prime interest, for they have already provoked considerable further debate. Fundamental provisions hold that 22 Bell system operating companies must be severed from AT&T. Thenceforth, they will be severely limited, in that they will be barred from provision of any interexchange or long-distance telecommunications services, customer premise equipment, and unregulated or value-added information services. AT&T, in contrast, will retain much of its vertically integrated structure, most especially Western Electric and Bell Labs—a fact not lost on foreign observers—and will be allowed to enter any enhanced services or equipment markets through an unregulated "Baby Bell" subsidiary. Control over the Yellow Pages—a two billion dollar money machine—passes to AT&T. As noted, intrastate long distance telephone service must be relinquished by the local operating companies and will also pass to AT&T. In this respect, as the California public utility commission asserts, the settlement "appears to anticipate a major redefinition of federal and state regulatory jurisdiction" as well ("California Assails ATT&T Accord, Says It Plans To Intervene," 1982: 12). Patents obtained after the settlement goes into effect would be exempted from the previous required licensing provision, to become the exclusive fruit of AT&T ("AT&T To Be Freed From Licensing Bell Labs Patents," 1982: 43). In short, the AT&T powerhouse is not jeopardized, but perhaps even strengthened.

The dust has not even begun to settle on the proposed settlement. New issues, most notably, the specific form of the more detailed plan to be submitted to Justice by AT&T, are being raised on a daily basis by concerned parties. Opportunities for further legal intervention by the Court or by Congress are widespread. Nevertheless, a few basic issues growing out of the proposed agreement cannot be ignored.

First, insofar as it overturns the 1956 consent decree and allows AT&T to provide unregulated, value-added services and merged telematics equipment, *the settlement is a major victory for both Bell and the largest corporate users*: "We do support the basic objectives of the settlement and believe that we can work within its framework," claims the International Communications Association ("More Comments On The AT&T-DOJ Settlement," 1982: 38). Representative Thomas J. Tauke, of the House Telecommunications subcommittee, thus concluded that "the winners are AT&T, some competitors and the big telephone service users" (in Block, "Congress Fears Consent Decree Will Lead To Higher Local Rates," 1982: 11).

While this chief goal has now seemingly been attained, however, the manner of its attainment is of serious concern to business users and competitors. There seem to be very few checks on AT&T's market power, while the vagueness of the settlement in several key respects leaves its concrete impact uncertain. In particular, it seems assured that—as a result of the settlement—the divested local operating companies will face substantial economic difficulties, if not even actual hardship. Long protected by the AT&T umbrella, different operating companies are in fact on very different economic footing. The stock market began immediately following the settlement to differentiate sharply between bonds issued by diverse Bell operating companies; the yield differential between the strongest and the weakest of these bonds increased sixfold between 8 January and 10 February 1982 ("Investor Confidence In AT&T Debt Shrivels, U.S. Panel Is Warned," 1982: 39). Thomas J. McGuire, executive vice president of Moody's Investors Service warned that the settlement is "unquestionably a negative development" for the credit quality of some local operating companies ("AT&T Accord Clouds Outlook On Phone Bonds," 1982: 8). AT&T will likely try to shift much of its vast corporate debt—at $47 billion, the largest corporate debt in the nation—and operating costs as well to the local telephone companies. The latter, simultaneously, will be stringently restricted from all but high-cost, low-profit local telephone service. The consequence can only be enormous rate hikes. The National Association of Regulatory Utility Commissioners, in fact, has already claimed that, should the agreement be approved as it stands, "(t)he potential for impact upon ratepayers would be . . . perhaps unparalleled in regulatory history"

("Progress, Many Problems Cloud AT&T-DOJ Settlement," 1982: 11). California's public utility commission is trying to intervene in the settlement, charging that "the cost of affordable local telephone service (is) at stake" ("California Assails AT&T Accord, Says It Plans To Intervene," 1982: 12). Robert M. Litke, Commissioner of General Services for New York City, informed a House subcommittee the city was concerned as a result of the agreement that its "elderly and poor . . . may find phones too costly in the years ahead" (Holsendolph, "Panel Told Of Phone Worries," 1982: D6). The time-hallowed goal of universal telephone service, in short, seems to have been placed in jeopardy.

Local operating companies can be threatened in this fashion because, for the largest business users, competitive alternatives are rapidly being developed. Through microwave, coaxial cable, and possibly cellular radio, new suppliers are seeking to provide users with enlarged data transmission capabilities, altogether bypassing older, inferior local operating company circuits (Hirsch, "Major Industry Battle Brewing Over Wideband Dems Planned by Tymnet, SBS, and Isacomm," 1981: 57; FCC Docket No. 79-188, RM-3247, RM-3497, "In The Matter Of Amendment Of Parts 2, 21, 74 and 94 Of The Commission's Rules To Allocate Spectrum At 18 GHz," Further Notice Of Proposed Rulemaking, 2 September 1981: 32-37). With numerous local facilities demanding the same sort of interconnection capacities currently available via satellite or microwave long-distance circuits, large firms are seeking—and beginning to find—alternatives. Indeed, under the proposed settlement AT&T *itself* would be permitted renewed entry into local services. Victor J. Toth, a communications attorney, thus states that AT&T likely will "attempt to provide some sort of direct connection to favored customer premises . . . completely bypassing the local exchanges" ("Bypassing Ma Bell . . . ," 1982: 29). The operating companies will be left with scant incentive to request politically sensitive rate increases with which to upgrade their facilities for digital data transmission. Why should they?—they are prohibited from value-added services anyway. Furthermore, the prospect of many independent telephone operating companies imposing different prices and conditions on local data transmission circuits may itself impel swift development of wideband communications links to bypass local lines (Blumenthal, "Industry's First Concern: Local Transmission," 1982: 10).

As the stripping of the operating companies is accomplished—unless prevented by those desirous of a continued competitive alternative in local service, including some business users—local telephone rate hikes, erosion of universal service, and deterioration in service quality appear increasingly possible. Vital social and economic and political issues will therefore have been decided, even if only by default. "What the FCC is doing," snapped erstwhile AT&T Chairman John deButts six years ago in congressional hearings, when deregulation was not nearly so far advanced as it is today, "might help some people, mostly big business" (Statement of John D. deButts, *Hearings On Competition In The Telecom-*

munications Industry, Before The Subcommittee On Communications Of The Committee On Interstate and Foreign Commerce, House of Representatives, 94th Congress 2d Session, 28, 29, 30 September 1976, Serial No. 94-129, USGPO 1977: 11). For most, deButts insisted, "it is going to mean significantly higher charges." "(W)hat your competition is going to do is to reduce rates for big business and increase rates for little people," deButts charged (Ibid.: 31).

DeButts was right. Yet he believed that Congress faced a far more fundamental issue even than this. "(W)hat is the basic aim of this country's telecommunications policy?" (Ibid.), the then-Chairman of AT&T inquired:

> Is it, as we in the industry have conceived it to be and the Communications Act appears to confirm, to promote the widest availability of high quality communications service to all of the people of the United States, or does that aim now yield to particularized interests of specialized classes of users . . . .

DeButts had to appeal to the "little people" simply because the big ones —business users—were arrayed against him. No matter if self-serving, however, his question remains profoundly valid today—for, in the great telecommunications debate it has for the most part been nimbly sidestepped, evaded or crisply ignored. While debate has swirled round the issue of "competition"—a smokescreen for business users' freedom of choice—the truly vital and encompassing issues raised by rapid integration of telematics into the conduct of our entire social life as a people have been mainly framed and addressed and, perhaps, resolved, in terms of private corporate strategy, and not of public good.

Yet the sweeping change that is upon us, and that has grown in scope and force over the entire postwar era, adds up to a transformation of the entire economic base of the United States. It is akin to shifts impelled by the railroad in the nineteenth century—or to those effected in conjunction with growth of electrical power industries between about 1880 and 1930. telecommunications industry expenditures for plant and equipment in 1978 accounted for over 11 percent of capital expenditures by all of U.S. industry—and this includes neither private-user outlays for private networks nor sums spent on computers (Bolter and Irwin, 1980: 37 and Chart 16). It is ultimately because this is indeed a general metamorphosis of the economy that it, like that economy historically, has assumed a private or deregulated character. What *societal* consequences will develop as a result? What will be the experience of the U.S. population as it is hurled pell mell toward an information age?

Broader still, what changes in the *world* economy and in *global* society and culture, will be enacted? For as we see that it has been the whole of U.S. big business that has fought to privatize telematics, we confront another momentous implication of the transformation that is upon us: it is, in its essence, transnational. Drawing force initially from the domestic U.S. context, privatization of telematics can only be considered properly within an international context. Again, this is owing to the nature of the corporate users at the center of what we have to this point analyzed purely in a domestic guise—for these, overwhelmingly, are transnational companies whose administrative, productive and distributional activities span the globe. It is therefore of the utmost importance that we now relinquish our domestic focus and turn attention to the dominant international features of the computerization of society in Part Two.

TWO

The U.S. Offensive In International Telematics

"These changes in domestic policies have important international spillover effects; and there is an increased need to reconcile the pro-competitive, open, flexible regulatory policies in the U.S. with national telecommunications policies overseas which are based on very different approaches to industry structure and the provision of service to users."
———— Robert R. Bruce, 1981

"The focus on domestic policy, however, should not lose sight of the fact that the welfare of our nation and the strength of its economy mandates a favorable balance of trade in the increasingly global marketplace for telecommunications products and services . . . . Surely it would disadvantage our country in foreign markets were the Bell System to be denied the opportunity to exercise its full strength in the face of competition with overseas suppliers . . ."
———— Charles L. Brown, 1981

The defining feature of the postwar epoch has been the sustained expansion of the United States economy across international borders. One measure of this growth, direct foreign investment, pinpoints the process: U.S. direct foreign investment grew from $7.2 billion in 1946 to $192.6 billion in 1979 (book value) (Wilkins, 1974: 329; Whichard, 1980: 16). Foreign branches of United States banks, paralleling the broader trend toward transnationalization of corporate activities, increased from 111 in 1955 to 779 in 1979; thousands of foreign business enterprises came under U.S. corporate control (Wilkins, 1974: 393, 374; U.S. Department Of Commerce, 1981: 502).

Administrative coordination and control of increasingly farflung corporate operations were similarly advanced through postwar years. Yet integration of dispersed worldwide activities—in manufacturing, marketing, shipments, storage, finance, insurance, advertising and so forth—demanded and today continues to require transformation of the means of communication. Varying by industry sector (banks may have different specific needs than energy companies), the pressing need is still everywhere broadly the same: How can communications be technically upgraded, modernized, and managed, the better to coordinate, integrate and control global productive facilities? Telecommunications has thus become a limiting factor in the corporate economy. "With the continuing demand in foreign trade," states Robert Barnwell of the Telecommunications Association, "international communications are of paramount importance to scores of industries" (Statement of Robert Barnwell, Before The Subcommittee On Communications, U.S. Senate 95th Congress 1st Session, 21, 22, March 1977, *Hearings On Domestic Telecommunications Common Carrier Policies:* 1047). Or, as the International Communications Association—the largest U.S. user group—puts it: "International telecommunications provides the pipeline which enables U.S. industry to extend its enterprise to the vast world markets" (FCC, "In The Matter Of Regulatory Policies Concerning Resale And Shared Use Of Common Carrier International Communications Servies," Comment of ICA 7 August 1980: 5). Capturing the essence of this relation, Harry Freeman, Senior Vice President at American Express, states that "Communications have become essential for the operations of all modern business—from steel and energy to airlines and banking. Modern business relies on open communications networks and the unrestricted flow of information for everything from processing of worldwide employment records and design work, to sales coordination and financial transactions" (Statement Of Harry L. Freeman, Before The Subcommittee Of The Committee On Government Operations, U. S. House Of Representatives, 97th Congress 1st Session, On H. R. 1957, *International Communications Reorganization Act Of 1981,* 31 March and 2 April 1981: 167). This vital dependence upon communications, made ever more acute by the rapid merger of communications with data processing and the host of applications spawned by their convergence, is part and parcel of the transnational corporation's long term need to have "*free access* and *free entry* into various national economies" (Ibid: 170, original emphasis). Corporate reliance upon telecommunications must, from the outset, be understood in terms of this broad and dynamic ideal of unfettered transnational corporate expansion.

One of the world's largest private, on-line data communications networks is operated by Texas Instruments (TI). The diversified microprocessor and electronics company has 50 major plant sites in 19 countries employing about 80,000 people. TI's system embraces 8,000 inquiry terminals and 140 distributed computers connected to the Corporate Information Center in Dallas. It is used for engineering, manufacturing (global production planning, manufacturing automation and control, tracking material in transit from site to site, and cost accounting on a product-by-product basis), financial planning, marketing and customer service (dedicated customer terminals tie into the TI network to facilitate instant maintenance of inventory), personnel management (managers in remote locations may rank the value of employees worldwide), electronic mail, electronic filing, word processing and teleconferencing. In use since 1971, the electronic mail system permitted about 25,000 transactions a day by 1979; satellite circuits interconnect the many different corporate sites. Further innovation is on the way. TI currently employs a scanning electron microscope in Singapore to view potentially defective integrated circuits; the image is digitalized and transmitted over the network for viewing in Texas. "It's still relatively slow," concedes L. C. Craig, information systems manager at TI, "but better than flying an engineer to Singapore" (L. C. Craig 1981: 213, 202–214 passim, "When TI Talks, the message moves fast," 1980: 103–104).

Other examples of generalized transnational corporate dependence upon telecommunications are not hard to find. Ford Motor Company links its Engineering Computer Center in Michigan to its European Automotive Operations headquarters in Dunton, England, and to its Research Center in Merkenich, Federal Republic of Germany. "These communications links not only allowed European Operations to cut costs by purchasing computer services directly from the North American data center, but permitted Ford engineering centers to share common computer-based tools and avoid costly duplication of efforts," the company explains ("Ford's Dial-Up Design Technique," 1981: 46–48). In its words, this worldwide communications network "made it possible to develop . . . the new world cars" (Ibid.: 46). Chase Manhattan Bank, on the other hand, must apply new information technology to coordinate its 106 branches, 25 representative offices and 75 subsidiaries doing business in over 100 countries. Through the company's corporate cash management service, balance and account information may now be retrieved in almost any country or currency in Western Europe. This service furnishes corporate financial officers with immediate information and allows them to transfer funds instantaneously from country to country to "insure full utilization of money," as a Chase representative claims. Such real-time updating of customer files allows instant retrieval of full descriptive reports, either through terminals located in international correspondent banks or through portable terminals at the customer's location. At Chase, generally speaking, "telecommunications has entered a period of explosive growth paralleling the rapid increase in information needs" associated with the bank's continuing expansion. (Letter from Kay Riddle, Vice President, Chase Manhattan Bank, to Richardson Preyer, Chairman of the Subcommittee on Government Information and Individual Rights, Committee On Government Operations, U.S. House of Representatives, 96th Congress 2d Session, *Hearings On International Data Flow*, 10, 13, 27 March, 21 April 1980: 739–742).

The same group of heavy corporate users of telematics which, as we have seen, has become a guiding force in an evolving domestic policy for U.S. telematics, *today likewise demands expanded and integrated telematics services in the international arena*. These major corporate users require global rebuilding of telecommunications network facilities just as they earlier pushed for domestic upgrading of the network.

Two points are relevant here. First, like its domestic counterpart, the dynamic international telecommunications scene has already been of long standing. The number of telephone messages to and from overseas has grown at a staggering and completely unexpected rate for *decades* (Table 6). Second, as is shown in regard to telex—a teletypewriter exchange service tailored to business—distribution of international traffic is asymmetrical (Table 7). The nations of Europe, which accounted for about 42 percent of total U.S. direct foreign investment in 1979, footed 48.9 percent of incoming and outgoing telex messages in that year (calculated from Table 27, "Analysis Of Overseas Telex Traffic By Country Or Point," in Federal Communications Commission, *Statistics Of Communications Common Carriers* Year Ended 31 December 1979, USGPO: 163–165). These same European countries accounted for about 41 percent of ingoing and outgoing U.S. telephone calls (calculated from Table 15, "Analysis

TABLE 6
Overseas Telephone Services
Calls

1929	30,323
1939	76,189
1949	852,865
1959	3,088,645
1969	20,800,303
1979	181,521,678

Source: Dan S. Fargo, in "Int'l telecom spending continues its climb," *Telephony*, February 23, 1981: 61, figures provided by Business Research-Long Lines Department, Long Lines Statistics AT&T 1980.

TABLE 7
U.S. International Telex Traffic: Top 25 Correspondents (To and Fro), 1979

Country	Number Telex Messages (000s) (Ingoing & Outgoing To and From U.S.)	% Total U.S. International Telex Messages (To and Fro)
United Kingdom	14,421	16.2
Japan	6,478	7.3
Federal Republic of Germany	5,412	6.1
France	4,603	5.2
Italy	3,692	4.1
Australia	3,008	3.4
Hong Kong	3,003	3.4
Switzerland	2,880	3.2
Netherlands	2,642	3.0
Brazil	2,599	2.9
Taiwan	2,582	2.9
Venezuela	2,034	2.3
South Korea	1,930	2.2
Belgium	1,707	1.9
Argentina	1,517	1.7
Spain	1,438	1.6
Sweden	1,356	1.5
South Africa	1,298	1.5
Singapore	1,252	1.4
Philippines	1,249	1.4
Puerto Rico	1,243	1.4
Saudi Arabia	1,166	1.3
Colombia	1,130	1.3
India	954	1.1
Israel	875	1.0
TOTAL 25	70,469	79.1
TOTAL U.S.	89,038	100.0

Source: Federal Communications Commission, *Statistics Of Communications Common Carriers,* year ended 31 December, 1979, Washington, D.C., USGPO: calculated from Table 27, 163–165.
Note: Mexico and Canada would both certainly rank in the top 25, were they not classed as domestic points.

Of Overseas Telephone Traffic By Country Or Point,'' Ibid.: 25–27). In contrast, the entire continent of Africa constituted only 1.1 percent of all telephone traffic to and from the United States in 1979 (Ibid.), while South America comprised but 11 percent of total telex traffic (Ibid., Table 27; 163–165). (It should be noted that these figures do not reflect traffic with Canada and only partially take into account traffic with Mexico, both of which have long been classed as "domestic points" for such purposes.)

If the international telecommunications industry is both dynamic and asymmetrical, however, the policy problems attending the direction and control of its further growth are unique, and vastly different from those we have encountered in the domestic sphere. In sum, most other nations have chosen to run telecommunications equipment and services through public government agencies—Ministries of Posts, Telephones and Telegraphs, or PTTs—rather than through private companies like AT&T. The key issues in the global reconstruction of telecommunications now underway return continually to this divergence. Shall the new networks be publicly owned and controlled, or shall they take the form of private enterprises? How shall they be regulated? What shall be the features of emerging rate structures? Which equipment manufacturers shall obtain what size pieces of the telematics pie—shall preference be granted to the particular domestic companies that have long enjoyed close collaborative association with the PTTs, or shall a "free market" philosophy prevail? Above all, whose interests shall predominate in policy formation for international telematics—those of particular nations or those of major private users whose global activities depend upon efficient and economical telecommunications? On whose terms shall the telematics transformation proceed?

In the United States, the center of the transnational corporate economy, the relation between domestic and international telematics market has, not surprisingly, become the cardinal axis of decisionmaking in this field. Instances of this increasingly close relationship abound and, in part, reflect the long standing attempt to use the huge domestic market as an important source of power in the world economy. "We can foresee, "declares one policymaker, "that as more consumers come to rely upon sophisticated and innovative services offered domestically, they will demand that these services be offered internationally as well" (Ahern, 1980: 359–360). In this respect, *deregulation within the United States can only be understood in light of an historically unfolding interaction between domestic and international policy*. Beyond the contest for position between AT&T, IBM, GTE, Xerox, and other major companies, we shall find, domestic deregulation plays a signal role internationally. One might say even that it signifies the hostaging of the domestic U.S. economy to transnational corporate interests. Thus, as Henry Geller, then chief of the National Telecommunications and Information Administration, told a congressional committee, "the most important thing we can do is to pass telecommunications reform domestically"—because

> If we subject the domestic industry to delays—to long administrative delays—before it can introduce services and products, if we regulate rates of return in competitive areas, if we do not allow competition to have full sway, we will be hurting ourselves very badly, and that will hurt us in competition against foreign industry in the United States and competition overseas.
>
> (Statement Of Henry Geller, Before The Subcommittee Of The Committee On Government Operations, House of Representatives 96th Congress 2d Session, *Hearings On International Data Flow*, 10, 13, 27 March; 21 April 1980: 306).

In using the pronoun "we," of course, Geller has identified the interests of transnational corporations with those of the population of the United States. Other observers, however, readily identify the former as the actual agents of change. Testifying before congress, William W. Betteridge of AT&T baldly declared that "there really is no longer a 'domestic market' separated from international dealings" (Statement Of William W. Betteridge, Subcommittee On Telecommunications, Committee On Energy and Commerce, U.S. House Of Representatives, *Status of Competition And Deregulation In The Telecommunications Industry,* 97th Congress 1st Session 28 May 1981: 366):

> Large customers increasingly expect to deal with their international telecommunications and data in a systematic, unified way. International systems solutions to communications needs are increasingly demanded (Ibid.).

Transnational users want transnational solutions. The "pro-competitive, open, flexible regulatory policies in the U.S." which have long been beholden to corporate users and which these users have tried so tenaciously to achieve, need to be "reconciled," according to Robert R. Bruce, erstwhile Counsel to the FCC, "with national telecommunications policies overseas which are based on very different approaches to industry structure and the provision of service to users" (Statement Of Robet R. Bruce, Before The Subcommittee On Information And Individual Rights Committee On Government Operations, U.S. House Of Representatives, Hearings On H.R. 1957, *International Communications Reorganization Act of 1981,* 97th Congress 1st Session, 31 March, 2 April 1981: 46). The very changes spearheaded by the Federal Communications Commission, that we have up till now considered purely in their domestic aspect—Carterfone, the Computer Inquiries, Specialized Common Carrier, Resale and Sharing, and on and on—have had profound repercussions overseas. That is, they increase demand by many of the same users who benefit from their results within the domestic market for similar services and new freedoms abroad. While these large transnational users demand "international systems solutions" to their communications needs, the U.S. policymaking apparatus is transformed into an appendage of the *global* economic force that this sector comprises. Finally, the task of completely harmonizing U.S. national communications policies with international economic policies—ceaseless and dynamic as it is—"is central to development of a U.S economic strategy" in an increasingly fragile and unstable world economy (Statement Of John Eger, Subcommittee On Government Information And Individual Rights, Committee On Government Operations, House Of Representatives 96th Congress 2d Session, 10, 13, 27 March, 21 April 1980, *Hearings On International Data Flow:* 185).

Part Two of this volume is dedicated to analysis of this multifaceted process through which U.S. telematics policies are bent to conform to the interests of transnational corporate users. The first section concentrates on attempts to revise policies so as to permit greater integration and expansion of telematics services available to transnational users. The second section focuses on a now-advanced attempt to export the (de)regulatory model adopted by the United States overseas. Throughout, the tensions and antagonisms that make the international telematics scene so different from the domestic situation—and which in fact decisively shape the domestic situation—are noted and analyzed.

POLICY FOR INTEGRATION AND EXPANSION OF INTERNATIONAL TELEMATICS

Rebuilding (in some cases simply building) the world's telecommunications infrastructure is proceeding at a torrid pace. Table 8 shows the progress of expenditures upon submarine cables over the course of two decades. Telephone construction expenditures of the top twenty-five nations are depicted in Table 9. The value of computers and related equipment worldwide is shown in Table 10. Throughout much of the developed and some of the developing world, as is evident from these vast sums committed to telematics, provision for the new infrastructure has become a top national priority. In France, for example, the PTT under Giscard d'Estaing was the nation's largest investor apart from the military ("The Technological Bubble," 1981: 14); the German PTT also is the nation's biggest investor (Burkert, 1981: 9).

Vast sums are being committed to modernization and building of national telematics infrastructures for several reasons. First, transnational corporate users are pushing to effect coherent advanced systems for international non–voice traffic—whose many national components alike must be efficient and compatible. Once this modernization process takes hold in one portion of the world economy, competition—for telematics equipment markets, for provision of value added services, and for the myriad processes and products now dependent upon telematics

TABLE 8
The World's Submarine Cables

Year	Circuit Miles in Operation	Total Investment (Then-Current U.S. $ Millions)	Capital Cost ($) Per Circuit Mile
1960	600,000	180	300
1965	4,800,000	672	140
1970	14,700,000	1077	73
1975	35,500,000	1715	43
1980 (est)	91,900,000	2977	32

Source: *1980 World's Submarine Telephone Cable Systems,* NTIA Contractor Reports. U.S. Department Of Commerce, May 1980, National Telecommunications and Information Administration—CR—80—6: 12–13.

TABLE 9
Telephone Construction Expenditures 1979—Top 25 Nations

Country	1979 Outlays ($ Millions)	1981 (Budgeted or Estimated)
United States	20,300	23,550
Japan	6,700	n/a (8,150 in '80)
France	5,700	n/a (5,600 in '80)
Federal Republic of Germany	3,100	4,579
United Kingdom	2,600	4,301
Canada	1,800	2,208
Spain	1,300	1,788
Australia	950	1,188
Austria	687	1,011
Brazil	597	1,696
South Korea	586	1,702
Netherlands	494	n/a
Belgium	479	457
Sweden	477	968
Norway	472	471
Switzerland	470	504
Taiwan	392	722
Argentina	366	825
South Africa	354	550
Venezuela	350	440
Mexico	321	663
Denmark	302	318
Finland	277	292
Ireland	154	412
Saudi Arabia	n/a	n/a
Top 25 Estimated Total	49,228	
WORLD TOTAL	63,102	

Source: Dan S. Fargo, "Int'l telecom spending continues its climb," *Telephony* 23 February 1981: 52–57.

TABLE 10
Number and Value of Computers and Related Equipment in Use

	1960		1978	
Region	Number	Value (U.S. $ Billions)	Number	Value ($ Billions)
United States	5,500	8.8	200,000	193.6
Western Europe	1,500	2.6	110,000	124.8
Japan	400	.5	45,000	33.6
Others	1,600	.8	95,000	72.0
WORLD TOTAL	9,000	12.7	450,000	424.0

Source: "**Japan takes aim at IBM's world,**" *World Business Weekly,* 20 April 1981: 30 (Diebold Europe, 1979).

for their production and delivery—engenders an overpowering impetus to rebuild elsewhere around the globe. Progressive deregulatory advance coupled with ever more numerous market applications within the largest domestic economy—that of the United States—has had precisely this effect. Other nations simply cannot afford not to compete. The awesome integrative capacity of information technology, with applications in everything from electronic mail and teleconferencing to computer-assisted design, data base information retrieval and inventory control, continues to enlarge the stakes of the game. It is this sweeping market expansion and redefinition which makes national information policy, and forceful commitment to modernization of the telematics infrastructure, so essential. Not merely today's markets but tomorrow's depend upon that commitment. If domestic interests cannot today gain enlarged markets in a specific subfield of telematics, they must nevertheless be prepared to ward off further incursions by foreign—usually American—groups who can. Thus the rebuilding process is at once a straightforward and an intricately complicated contest. On one side, how can telematics building and modernization within a given country be directed and channeled to maximize (and, where needed, balance) domestic economic interests both now and in the future? On the other hand, from transnational users' perspective, how can the same process be accomplished best to expand flexibility, private control and predictability for users themselves?

It is a conflict exacerbated by global recession. Today, economic protectionism, trade wars and heightened competition have reached a scale unsurpassed in forty years. Heeding the cries of domestic suppliers, U.S. authorities have subjected cars, shoes, steel and consumer electronics to varying measures of protection from lower-priced imports. In Western Europe, Japanese autos and electronic and household appliances have reached a level of market penetration which now raises similar protectionist impulses. Japan in turn now charges that U.S. barriers of various kinds are impeding free trading (Zahn, 1980: 68; Keatley, 1981: 27; Treadwell, 1981: 4B; "Japan Intends to Go on the Offensive, Ask U.S. To Drop its Own Barriers to Trade," 1981: 33). Huge trade deficits—expected to total about $28 billion for the U.S. in 1981, and predicted by some to exceed $50 billion by 1982—foster giant instabilities ("Trade deficits spur the case for protection," 1981: 48), as the political costs of the economic malaise become more sharply defined. Economic rivalries between postwar allies—the U.S., Japan, Canada and Western Europe—likely may become "the priority issue of U.S. foreign policy," while the basic issue of the emerging era will be "who will export unemployment to whom" (Wallerstein, 1980: 125).

For the United States' economic leaders, what Harry L. Freeman of American Express calls "the pressures of the new protectionism" hold special perils (Statement Of Harry L. Freeman, Before The Subcommittee On Government Information And Individual Rights Of The Committee On Government Operations, U.S. House of Representatives, *International Communications Reorganization Act Of 1981*, 97th Congress 1st Session On H.R. 1957, 31 March and 2 April

1981: 168). This is so owing both to the scale and the quality of U.S. transnational economic interests. The sheer size of U.S. direct foreign investment (measured at book value, over $190 billion in 1979) is far greater than the equivalent stake held by any other nation's companies. The U.S. interest in cultivating a world telematics infrastructure maximally responsive to transnational users is therefore also proportionately the greatest. The startling growth of U.S. services—the so-called "invisible trade"—with their heavy reliance upon advanced telematics capabilities, again increases the ultimate importance of the latter to U.S. economic needs. Services suppliers including advertising agencies, banks, public relations firms, accountants, insurance companies, computer software houses, have largely piggybacked themselves atop transnational corporate growth itself. Eleven of the top twelve advertising agencies are today U.S.-based ("Foreign Agency Income," 1981 Section 2: 1); roughly two-thirds of a sample of 200 of the largest transnational companies were found to be audited by the "Big-Eight" U.S. accounting firms ("International Accounting Firms Consolidate Worldwide," 1980: 9). The bouyant services sector accounted for almost a third of total U.S. exports in 1979, about $35 billion (Kirchner, "Administration to Push Services Trade Overseas," 1981: 84). Because services tend to be laden with a heavy information component they tend also to be at the forefront of telematics applications; because, increasingly, services themselves are provided transnationally, telematics capabilities must be found to permit such international activity. Assurance of the availability of a raft of services, from cash management to production scheduling, "is thus of primary concern to all multinational corporations, regardless of the product they provide" ("A Preliminary Review of Barriers to Trade in Telecommunications, Data Processing, and Information Services and Transborder Data Flows; Motivations for Imposing Barriers, Trade Implications and Possible Approaches to Resolution," Draft dated 3 September 1980, Office of the United States Trade Representative: 2).

A vivid and atypical glimpse of what acute dependence upon telematics-based information services may lead transnational users to do is furnished by the case of an approved International Telecommunication Union recommendation for the technical design of packet switched networks—the X.25 series of protocols. X.25 comprises a technical standard; it is said to have been adopted in Europe "as a device to prevent IBM from establishing their system network architecture as a widespread standard; this allowed European companies which do not make entire systems to compete on the basis of compatible components" (Sirbu 1981: 194). In other words, X.25 was eagerly embraced by national authorities whose domestic telematics capabilities were no match for IBM's, and who hoped to utilize X.25—the form of which differed substantively from IBM's own design—as a means of gaining leverage in the telematics rebuilding process. Instead of being locked in to IBM's vision of their future, the Europeans thus erected an incompatible technical standard that would allow various European companies to participate in supplying components of the new infrastructure. All this, it should be stressed, is routine, as technical standards have always been a neglected but critical facet of economic rivalries in the communications field (e.g., Crane 1979).

What *is* unusual is that U.S. transnational corporations, in the face of the Europeans' intransigence, seem to have endorsed the X.25 standard. In a matter of a few months, Manufacturers Hanover, Chase Manhattan and Bank of America announced their support for X.25, the U.S. Federal Reserve bruited the idea of acceptance, and the Federal Government endorsed an X.25-based interim standard for its National Communications System ("Manufacturers Hanover Trust Integrated Telex/Private Net," 1980: 61; Schultz, 1980: 1, 5; "Federal Government Endorses Version Of X.25, Seeks Comments," 1980: 4). Bank of America, which on a busy day passes $20 billion in assets through its worldwide network, "cannot stall its expansion planning until IBM gives its blessing to a de facto international standard," claims one report (Schultz,1980: 5). Yet more unusual, large users' demands found their mark even over the interests of IBM, with its tremendous market share of the world's computer base. In Summer, 1981, IBM announced its decision to support the X.25 standard within the United States. According to one analyst, this decisive move in favor of the Europeans took place only because "IBM caved in to the clamor of customers . . . . IBM was really forced to support it" (Hoard, "IBM to Support X.25, X.21 in U.S.," 1981: 1, 6). Disputes over X.25 are, however, merely a part of the far broader and deeper contest that is telematics today.

Users and The International Information War

Global economic slump combined with sudden recognition of the centrality of telematics to the evolving world economy have ushered in what one participant calls an "international information war" (Eger, 1981: 103–119). Another erstwhile government spokesman declares that the critical fields of telecommunications and information policy must be prepared "for the coming trade wars" (Henry Geller in Kass, 1980: 80). Such language is unparalleled in the postwar period for its frank recognition of opposed economic interests held by European, Japanese, and United States companies.

But if the growing international information war is a battle between transnational firms, it holds two other equally important aspects. First, it has been largely developments within the dynamic United States market—combining technical advance, market demand, and gradually restructured regulatory policy—that have compelled users throughout the world to insist on similar benefits. Such liberalized domestic policies, accords onetime General Counsel to the FCC Robert R. Bruce, "have important spillover effects" (Statement Of Robert R. Bruce, Before the Subcommittee on Information and Individual Rights Committee on Government Operations, U.S. House of Representatives, 97th Congress 1st Session 31 March, 2 April 1981, *International Communications Reorganization Act of 1981:* 46). Second, because of the monopoly grip over telecommunications held by national authorities abroad, such "spillover effects" must inevitably confront directly the efforts of foreign governments to provide maximal benefit to their own constituents. Transnational—dominantly, U.S. transnational—users must therefore carve out worldwide telematics systems in an environment of increasingly fractionated national economic and political involvements. The basic question for users thus becomes (Ibid.: 47): "in a global economy with U.S. multinational corporations whose activities transcend national borders, what mechanisms will assure that these users' access to diverse and innovative telecommunications and information services is not unreasonably impeded?"

Government authorities, we shall find, have not been without assistance in formulating answers to this key question. Especially have they found support from users and user groups, whose international role is as significant as is their domestic policymaking presence. Historically cornered "between telecommunications authorities, manufacturers and labor unions" in the international business arena, users have now begun to emerge as "a fourth partner, commanding increasing attention and weight" (O'Neill, 1981: 56).

Deutsche Telecom eV, an association whose members must spend at least DM 1 million annually on telecommunications, and who must not be manufacturers or distributors of telecommunications products, founded in 1978, recently comprised 78 full time and 846 indirect members. Sectors represented by members are: electrical (29 percent), administrative organizations (19 percent), steel and engineering and automotive (14 percent), banking, finance and insurance (14 percent), chemical industries (7 percent), transport (7 percent), and construction (3 percent), and a few others. Scarcely 21 percent of member companies are of merely national scope; 48 percent are national with international subsidiaries, while 27 percent are subsidiaries of foreign transnationals ("User group survey—Deutsche Telecom eV, West Germany," 1979: 228). The presiding chairman of DTeV is manager of administrative services for the European office of John Deere, in 1980, the 66th largest U.S. transnational corporation ("The 100 largest U.S. multinationals," 1981: 92).

The Belgian Telecommunications Managers Association (TMAB), formed in 1977, represents 52 members drawn primarily "from international companies." A very small number of Belgian companies without transnational components is represented. The three main sectors comprising the membership are manufacturing (petrochemicals, textiles and food), finance and services ("User group survey—Telecommunications Managers Association, Belgium," 1979: 63).

The Swiss Association of Telecommunications Users (ASUT) encompasses 103 members of which about one-third are based in French Switzerland. As its chairman declares, "multinational firms are well represented." Banks, as "is only right and logical," are "strongly represented." Insurance companies, chemical firms, import-export houses, the police, publishers, travel agents, machine manufacturers, retail chains, colleges and mail order companies are also encompassed ("Swiss Association of Telecommunications Users," 1980: 41).

In the United Kingdom, the Telecommunications Managers Association embraces 122 organizations in banking, insurance, oil, chemicals, manufacturing, automobiles, aircraft, electrical equipment, transport, wholesale and retail companies, the press, local authorities and government organizations. Scarcely half (52 percent) of its members are national concerns ("Telecommunications Managers Association, UK," 1980: 40).

Both the diversity of such organizations—their inclusion of private and public sector groups—and their penetration by transnational firms, are of pressing importance. To the extent that these user groups encompass U.S. transnationals, which will certainly differ in each national grouping, they doubtless facilitate extension of U.S. users' reach into the heartlands of domestic telematics policy in other nations. Of further aid to U.S. transnationals' leverage are the boundlessly dense and virtually unprobed network of joint ventures between U.S. and foreign companies, private contracts for equipment and services by U.S. transnationals with patronage largesse to dispense, and financing and credit arrangements provided by U.S. banks to other nations and foreign companies. According to Herman (1981: 234), a 1975 survey of the 100 largest U.S. industrials revealed that 67 had joint ventures with one or more foreign companies, encompassing 56 separate foreign firms in fifteen countries, for a total of not less than 1,131 joint subsidiaries. Yet, he states, "(t)his is a gross underestimation, resulting from the severely limited disclosure requirements governing joint ventures." Direct intercorporate ownership is also apparently "extensive" internationally; a 1975 tabulation found that stockholdings by the 100 largest U.S. corporate investors in France included 819 individual holdings of one percent or more (Ibid.). From a different perspective, the French Ministry of Industry calculates that foreign penetration of the French economy varies by industry subsector from about 20 percent for machine tools to 81 percent for office machinery, computers and data processing— yielding a combined index of foreign economic ownership in the vicinity of 23 percent! ("Foreign ownership in France," 1981: 15). (Only a fraction of this total, of course, is directly attributable to U.S. direct foreign investment.) These findings suggest the scope of possible influence attained by transnational economic interests over various national economies.

Despite the gargantuan holdings of U.S. companies overseas, and the undoubted influence which these interests confer, there are several reasons why foreign and international user groups neither should nor can be equated strictly with U.S. transnational corporate interests. First, foreign investment within the United States has been growing rapidly of late, for a variety of reasons having to do with access to markets, materials, and political and economic stability. The

100 largest foreign investors in the U.S. commanded assets in 1980 well in excess of $170 billion ("New faces in an age of megadeals," 1981: 83). And this foreign investment is overwhelmingly dominated by Western European, Japanese and Canadian firms; 27 of the top 100 investments involved U.K. companies, making Britain the most important home base for foreign investors; 18 involve Canadian corporations; 16 reflect German holdings. These figures reflect direct ownership, as well as partial interests, and therefore are not comparable with the figures for U.S. foreign investment, above. However, a recent tally of foreign direct investment in the U.S. (net book value of equity and outstanding loans to U.S. affiliates by foreign direct investors) shows that of a total of $52.3 billion in 1979, European nations and Canada held almost $43 billion (Chung and Fouch 1980: 38). Indeed, the top eight countries comprised almost 90 percent of total foreign direct investment in the U.S.—in descending order, these were: Netherlands, U.K., Canada, West Germany, Japan, Netherlands Antilles, Switzerland and France (Ibid.). A single example must suffice to illustrate why this reciprocal direct foreign investment is germane to our discussion.

Societe Generale, nationalized some time ago by the French government, with assets of about $65 billion, is the world's seventh largest commercial bank. Starting with a restricted agency operation in the early 1940s, Societe Generale gradually expanded its U.S. activities until, in 1978, a decision was taken to develop "a large wholesale and retail operation" in New York. "We had to find and install the right computer system which would be the heart of our entire operation," declared a bank official ("Automated System Speeds Operations For International Bank's U.S. Branch," 1980: 38). At this juncture Societe Generale becomes indistinguishable from U.S. transnationals. "We have been observing for some time an increasing interest among non-U.S. banks to provide cash management services and it will only be a matter of time before competition in this field becomes truly international," adds Robert Walker of Continental Illinois National Bank and Trust (Statement Of Robert E. L. Walker, *International Data Flow* 1980.: 132). "It would appear, therefore, that a free information flow is as important to banks of other nationalities as it is to American banks or for that matter to multinational corporations worldwide," Walker summarized (Ibid.).

Developments within the United States domestic market therefore reach far beyond it. When companies can harness newly efficient administrative tools and productive processes by means of telematics, their incorporation races toward a new global business standard. Because nations exist within a world market economy, competitive advance in one part of the system compels analogous growth elsewhere. The number of foreign bank offices in the U.S. is growing rapidly—it reached 345 by 1979—implying that for this strategic sector competing foreign banks must seek to advance their own competitive advantages by pressing for advanced telematics capabilities at home (U.S. Department Of Commerce, *1981 U.S. Industrial Outlook*: 502). This certainly does *not* mean that foreign and U.S. based transnationals are indistinguishable in their strategy and tactics for achieving advanced telematics capabilities; merely, it means that they harbor an identical demand for upgraded and modernized telematics infrastructures. How this mix of common and opposed interests translates into concrete action, particularly by or through international user groups, is rarely obvious.

Two other qualifying factors constrain the efforts and activities of foreign and international user groups—and relate to the place and powers of PTTs within a newly volatile world economy. We have seen that these ministries have become major economic forces within their own domestic economies. This is so because the rebuilding of the telematics infrastructure is very expensive, but it must also inevitably pose the issue of government intervention into, and even control over, the economy. If the state itself is not to assume operating authority over an ever-widening sphere to meet the integrative potential and competitive necessity of telematics innovation in a world economy, then its current monopoly grip must be relaxed so as to allow private firms to do so. Concrete applications of telematics, as we found in analyzing the U.S. domestic scene, lead into areas far removed from the business of conventional telecommunications. This implies that the very process by which PTT authorities embrace telematics applications in their attempts to keep domestic equipment and services suppliers competitively alive will force them into other areas—banking and financial services, publishing, computer services, and so forth. It perhaps need not be added that the prospect of such dramatically enlarged PTT intervention into the economy raises the hackles of both domestic and diverse transnational business competitors.

Expanded PTT economic involvement occurs at a time of rapidly mounting economic competition internationally, as we have seen, and this translates into a new movement to enhance *private* control over areas which, due to the integrative trend of telematics, increasingly overlap with PTT operating authority. Proctor and Gamble is reported to be leading a campaign which may come to involve many other U.S. and European advertisers, "to increase pressure on European governments" so as to obtain more commercial TV time ("'P&G Heads drive for more Europe TV time," 1981: 1, 102). Other companies reputedly active in this effort are Colgate-Palmolive, Henkel, Unilever and Richardson-Merrell. The drive is aimed not merely at currently perceived scarcities of advertising channels, but also at ensuring that future expansion is unhindered by resistance to demands for more television advertising. Anthony Garrett, Proctor and Gamble's European vice president for advertising and purchasing, asserts (Ibid.: 102): "(I)t is improbable that a consumer product manufacturer, in business throughout Europe will plan major new investments on less than total European evaluation of the opportunity, chance of success and probable return." Marketing approaches must be devised, Garrett believes, that are "no longer isolated by distance and culture. In today's European market the cost and profit structure of our business will not tolerate the waste and duplication of unnecessary differences from country to country" (Ibid.). "Waste and duplication" are of course easier to perceive when direct broadcast satellites make feasible blanket coverage of many European nations at once. In any case Garrett declares, the issue of more commercial advertising time "is perhaps the most serious issue for the advertising business in Europe" (Ibid.:1). Five key points mark the hoped-for expansion by transnational advertisers across Europe (Ibid.: 102): permission for advertising on publicly controlled television; expansion of advertising time on existing channels; elimination of allocatory discrimination in favor of smaller advertisers; support for new advertising media—cable TV, satellites, video-recorders and videotex; and creation of "an environment for new electronic media which favors advertising." The general context in which transnational users of telematics are active today is ripe for a renewed attack on government controls over the economy which prohibit or restrict transnational economic integration on an epic scale.

All across Western Europe, major users and their newly formed associations are pressing to achieve a familiar goal: liberalized PTT policies for privately supplied foreign attachments—that is, specialized equipment attached to the PTT network—and provision of innovative and advanced services in private as well as public hands, particularly in the data communications field. When asked how satisfied members were with existing PTT conditions of service, Ernst Weiss, chairman of the West German users group replied: "If our members were satisfied, the DTeV would not exist" ("User Group Survey—Deutsche Telecom eV, West Germany," 1979: 229).

In the Netherlands, a study commissioned by a business group "strongly questions the role of PTT"—its main thrust is "for competitiveness and against government involvement in the information industry." The study—by Arthur D. Little, the U.S.-based consulting firm—attacks the organization and rate structure of the PTT. As "one combined organization active in financial, telecommunication and postal services, of which the latter two are a monopoly"—"Who is subsidizing whom?

> Are tariffs based on cost? Is the competition with banks in financial services fair? Would a divestiture encourage innovation in the separate activities? ("Dutch Publishers Sponsor Report Attacking PTT," 1980: 21).

In France, the Association Francaise des Utilisateurs du Telephone et des Telecommunications (AFUTT), founded in the late 1960s, has actively developed and cultivated a user agenda. Since the early 1970s, AFUTT has been an important contributor to the development of a national priority for upgrading the French telephone network. AFUTT has obtained from the (previous) Government provision for a consultative commission, bringing together PTT executives and user representatives at frequent intervals. Steps were taken under the administration of Giscard d'Estaing toward "a more adaptive and market-like policy" by the PTT in response to complaints from "large users and multinational corporations" (Chamoux, 1981: 22). These encompassed cheaper international telecommunications rates for EEC and U.S. circuits; publication of PTT annual accounts, "which will make possible a cost-effectiveness analysis of the tariff structure"; modification of PTT procurement policies to engender competition, in the hope of enhancing export capabilities by "liberalization in the marketing of a large number of telematics and telecommunications terminal products" compatible with PTT standards; and nurturing of closer cooperation between users (especially from banking) and equipment manufacturers. According to its 1980 Annual Report, the French PTT ministry believes that a PTT monopoly over telecommunications "must not represent a hindrance to the generalized use of terminals by the public," and, accordingly, favors a move toward systemic national integration of telematics throughout French society by facilitating closer cooperation between users and public authorities. Domestic goals for telematics development are placed squarely within an unfolding international context so that French exports are given top priority (Direction Des Affaires Industrielles Et Internationales, Direction Generale Des Telecommunications, 1980: 10,26,3–4).

The impact of the election of Francois Mitterand, with his extensive plans to nationalize major banks and five industrial groups, will surely influence the context of users' demands—perhaps dramatically. "(N)ationalization," states Mitterand, "is a weapon to protect France's production apparatus" from coming even further "under international control" ("Mitterand: Why Nationalization Will Work," 1981: 27). It is part of a generalized attempt "to win back the market at home" (Ibid.). The new Government has confirmed predictions that it would increase state backing for data processing and telecommunications; industrial strategy in telecommunications and telematics will "be maintained and reinforced" ("French Government States DP Plans," 1981: 29). Yet AFUTT is said to have "obtained certain commitments from the main candidates in the French Presidential elections," such that "definite efforts will be made to upgrade the quality of telecommunications in France, change pricing policies and approve the laws and regulations governing the relationship between users and the PTT." Future goals are "to increase the orientation of the PTT toward meeting user requirements in the development of new services and equipment," and "attractive prices and policies" are likewise of significance (O'Neill, 1981: 57–58). Although it cannot be predicted what course French telematics will follow, it seems evident that users are mobilized and poised to demand further favors.

In the Federal Republic of Germany, DTeV aims to cultivate a "steady and reliable communication link" between users, equipment manufacturers and the PTT ("User group survey—Deutsche Telecom eV, West Germany,"1979: 228–229). DTeV identifies and tries to solve economic, organizational and technical problems common to private user networks over which the PTT has no mandate. Members feel that the German PTT has neglected the interests of business telecommunications "in favour of public or residential interests," even though the former supply the larger and more rapidly growing users (Ibid.: 229). "It is our members' concern," asserts the presiding chairman of DTeV, that the PTT does not realize the importance of telecommunication as an economic factor in modern business. West Germany is, like most European countries, an export-oriented industrial country. Therefore, we must try to remain compatible on the world market and telecommunication is to modern industry, next to labour and materials, the most important expenditure.

Members hope that by means of "a strong user organization" the restrictive telecommunications policies characteristic of close bilateral exchange between PTT and equipment manufacturers may "evolve into a much fairer three-way interchange" (Ibid.).

Should it occur, such an advance would center on introducing improvements in data transmission services. All-digital transmission, introduction of value-added services like viewdata and electronic mail, permission to utilize a wide range of customer equipment and more careful cost-based pricing are alike encompassed by the German users' agenda (Ibid.: 229–230). The push for cost-based pricing, it should be stressed, goes to the center of the Government's most basic structure and policy (Ibid.: 230):

> We do not object to the fact that a service of the PTT is subsidizing another service, as long as the PTT telecommunications structure in itself is cost-covering. We are, however, strongly against the fact that revenues from PTT telecommunication tariffs are used to subsidize other governmental services, such as the unprofitable mail service, street construction, or even military services.
>
> The governmental demand on the PTT administration to surrender its profits leads to excessive tariff pressures on potentially profitable business telecommunications, which in turn results in unfair restraints to the business economy.

Such restraints, it may be added, become all the more onerous for business users in a climate of economic downturn. As viewed by DTeV, therefore, obstructions placed before users by PTT administrators may only be identified and removed during a process of "debating with the PTT its attitude to its monopoly rights" (Ibid.).

That debate has now been joined. The German Monopoly Commission, a standing advisory group, has published a report on a study of the Deutsche Bundespost, the German PTT. The Commission's findings, according to its head, are based on the premise "that in the context of the market economy of the Federal Republic a functioning competition process should have basic priority over any kind of public planning or regulation" (Burkert, 1981: 9). The Commission concedes in its report that the PTT monopoly over the national telecommunication network should be retained, at least for the time being, but claims that the PTT should admit special networks more liberally and should generally permit resale of private lines (Ibid.). With regard to all terminal equipment other than standard telephones, the Commission proposes "a free and unrestricted private market"—so free, in fact, that the PTT should be altogether excluded from it. Finally, the Commission would like to see the national market opened up more to foreign competition (Ibid.). Although the outcome of this growing debate cannot be predicted, once again it may be said that users' policy role has dramatically expanded.

In Switzerland, the Association Suisse des Usagers de Telecommunications (ASUT) has followed the frequent tactic of obtaining representation within the general management of the Swiss PTT. This liaison provides a useful point from which to pursue diverse user objectives. ASUT has already relied upon its foothold within the Swiss PTT to oppose "completely" plans by the latter to assume control over further development of word processors, hoping rather to keep the word processor "an entirely private piece of equipment" ("Swiss Association of Telecommunications Users," 1980: 43). "One of the main reasons for the creation of the association," we are told, "was to be able to influence the (PTT monopoly—DS) policy on equipment"; according to ASUT's chairman, M. Salim, "a heartening degree of collaboration has developed" (Ibid.: 42). Asked whether ASUT wished for policies allowing each user to acquire equipment as demanded from different suppliers, Salim responded (Ibid.): "This is our dearest wish . . . . It is mainly the large user who must be in a position to choose freely the equipment he needs."

Most illuminating, however, were Salim's comments on the increasingly vital international constraints placed on domestic telematics policy. ASUT felt duty-bound to inform the Swiss PTT that "there are limitations, even for an administration that has a monopoly over telecommunications" (Ibid.: 41). "We explained," Salim offered, "that multinational firms, the important telecommunications users, can transfer their communications centres without hindrance so that a country which offers more advantageous tariffs will be the beneficiary." For example (Ibid.),

> a firm which has its European branch in Geneva must order a certain number of rented lines for its calls to Europe, the Far East and Africa. For the PTT this represents a lucrative source of revenue. But such a technical centre could easily be transferred to the UK or France, which would enable considerable reductions in telecommunications costs.

"Concrete examples," according to Salim, verified that such intracorporate decisions are very much in the air. In the current period of economic hardship and growing unemployment throughout much of Western Europe, rising tensions between Common Market countries equally intent on preserving jobs and expanding profits domestically make it more likely that protectionist tactics will be closely considered in relations between European countries ("The EC nations may turn on each other," 1981: 62). Widening differences among European nations may, of course, effectively be played off against each other by transnational corporate interests, as Salim suggests is already taking place.

Which brings us to 10 Downing Street, where British telematics policies are drawn up and where—of all Europe—the cause of a "non-monopolistic" environment has been most extensively and enthusiastically endorsed.

Political and economic links between the United States, the core of the transnational corporate economy, and the United Kingdom, have of course been intimate over many years. The U.K. is the recipient of about 12 percent of all U.S. direct foreign investment (the only nation with a higher total is Canada); in turn, British direct foreign investment in the United States is second only to that of the Netherlands—accounting for about 18 percent ($9.4 billion) in 1979 (Chung and Fouch, 1980: 40). These, too, are ties that bind—at least in terms of telecommunications circuits. Postwar Britain has attracted "more and more transatlantic lines for large Euro-American corporations" (Chamoux, 1981: 22). According to British Telecom International, fully 60 percent of U.S. leased transatlantic data lines currently terminate in the United Kingdom (Schultz, "Integrated Teletex Network Predicted for UK by '82," 1981: 12). After the United States, Great Britain maintains the busiest Intelsat earth stations. The U.K. therefore quite clearly represents a key node in transnational telematics. Put crudely, if in Britain provision of terminal equipment and advanced telematics services is "privatized"—to use the term thrown forward in current British discussion of the subject—transnational corporate leverage against national protectionist strategies across all of Europe will be profoundly benefited.

Globally, "monopolistic arrangements"—that is, government PTTs —"are now under enormous pressure" ("Toward the 'wired society'," 1981: 29) to liberalize provisions governing employment of advanced services for business users. For several reasons, having to do not only with the unique relations between the U.K. and the United States, but also with the prodigious economic decline down which Britain is sliding, and, particularly, the apparent failure of its chief computer manufacturer ICL to compete effectively with foreign rivals (Rubiner, 1981: F9), the U.K. has proved especially vulnerable to such pressure. Britain indeed seems to have touched off a race to be second in worldwide telematics—after the United States—insofar as engendering a climate most conducive to corporate user interests is concerned. At the very least, Britain is testing the wind, looking by turns first to the United States and then to continental Europe for clues as to the most favorable national telematics course to set. Signs today point to progressive disengagement of the U.K. from continental Europe—polls claim that barely 30 percent of Britons favor retaining Common Market membership (Lewis, 1981: F2)—in favor of a stance attaching Britain yet more closely to U.S.-based transnational companies. Thus Dr. Janet Morgan, a member of the select advisory group to the British Prime Minister and Cabinet on matters falling outside normal government ministry jurisdiction, and an advisor to American Express, testified before a U.S. congressional committee that: "It is in important respects misleading to think of West European countries as a bloc, all displaying similar attitudes to communications questions and, accordingly, formulating domestic and foreign policies that closely resemble one another" (Statement Of Dr. Janet Morgan Hearings Before The Subcommittee On Information And Individual Rights Committee On Government Operations House Of Representatives 97th Congress 1st Session On. H.R. 1957, *International Communications Reorganization Act Of 1981*, 31 March and 2 April 1981 USGPO 1981: 28). When asked by Congressman English about the future plans of Common Market countries with respect to regulation of telematics, Morgan replied (Ibid.: 22):

> Perhaps one has the specific problem of a number of community countries, all now very well-aware that this is big business and is important, looking over their shoulders at each other. But at the same time, they are looking forward and outward at the United States and Japan. There is very interesting tension between the two perspectives.

Such "tension" may be important in explaining why Britain, according to Dr. Morgan, "has not replied formally to the Davignon plan for a European Community strategy in these matters" (Ibid.: 40). The latter is an initiative pursuing creation of a pan-European capability in microelectronics, liberalization of PTT procurement *within* the Community, and similar liberalization of the terms set for supply and sale of network foreign attachments. "It is instructive," claimed Dr. Morgan (Ibid.: 39), "to compare the current British approach to these matters with that of other European members of the European Conference of Postal and Telecommunications (CEPT)." What has been the nature of this approach?

The newly passed British Telecommunications Act allows competitive supply of terminal equipment for attachment to communications networks and considerable liberalization in supply of previously monopolized telematics services, including private telecommunications networks and, to some as yet uncertain degree, resale of PTT-provided circuits with value added ("UK: private networks to be allowed," 1981: 5). The Minister of Information Technology, Kenneth Baker, asserts that the new Act would "allow the private sector much greater freedom to use British Telecom's inland network . . ." (Ibid.). Many telecommunications users, writes one recent analyst (Peltu 1982: 80) hope the U.K. initiative "will spread like wildfire," while U.S. equipment manufacturers look at the newly liberalized U.K. market as "a stepping-stone to the rest of Europe" (Ibid. quoting Roger Camrass: 86).

The British Telecommunications Act and the policy initiatives which led up to it did not exactly proceed apart from user interests and activities. The British users' group, called the Telecommunications Managers Association (TMA), was centrally involved in passage of the bill through Parliament; according to one account, TMA "set up a subcommittee which influenced the Bill in the Parliamentary Committee stage" ("Telecommunications Managers Association, UK,"

1980: 40–41; O'Neill, 1981: 59). TMA's objectives are not noticeably at variance with those of major users elsewhere. Independent supply of customer equipment is strongly encouraged; sharing of leased circuits, now but rarely permitted, is favored as an option; cross-subsidies of various kinds are challenged; and, generally, "the role of PTTs and their monopoly powers should be investigated" ("Telecommunications Managers Association, UK,"1980: 40–41). The legislated separation of British Telecom from the previously unitary British Post Office which held sway over both telecommunications and postal services, was itself apparently accomplished so as "to isolate the real costs and subsidies associated wth each of these different activities"—and therefore may be counted as another user achievement (Statement Of Robert R. Bruce, *International Communications Reorganization Act Of 1981:* 49). Yet another users group, the Tele-Communications Users Association, worked "behind the scenes to encourage liberalisation of the British telecommunication network" (O'Neill 1981: 60), and with the Post Office Engineering Union called for restrictions on investment in British Telecom by private interests to be lifted. Today, British Telecom is limited to financing future capital expansion out of internal funds and revenues. Britain's capacity to create a modernized telecommunications network "able to meet the demands of present customers and be compatible with new technology" is said to be compromised by this constraint, as modernization will require investment of at least two billion pounds annually for the next five years ("Users, Unions call for more investment in British Telecom," 1981: 38–39). Significant about the argument employed by this coalition in favor of opening British Telecom to private investment is that "(i)f investment in telecommunications in the U.K. network is not brought into line with that of other countries. . .British business will suffer an ever increasing handicap due to poor communications . . . . At the same time, the British electronics industry will be deprived of an adequate home market on which to base sales of new equipment" (Ibid.). It goes without saying that not just British business, but all business located in Britain will share the burden of prospectively inferior telecommunications. With private investment in British Telecom, however, inevitably a degree of power over policy decisions would be conferred on private interests, including possibly those of users. Indeed not only in this case but also throughout its overall design, the changing British telematics environment will tantalize and probably promote foreign interests. According to an authority who questions the application of the U.S. domestic model to the markedly different British scene, "The Government should be under no illusions about American hopes and intentions as we appear to move towards a more competitive environment for telecommunications in this country" (Darlington, 1981: 46). U.S. telematics equipment companies as well as users expect "that the British decision to open up its telecommunications equipment markets will also open up the rest of Europe" (Ibid.).

Yet further relaxation of public control gives promise of taking place. Agreeing that the very profitable international business of British Telecom "might be endangered by progress towards liberalization," a report which supported the campaign for the British Telecommunications Act found nonetheless that "allowing resale at home and aiming towards its international adoption is a feasible strategy which promises advantages both to British Telecom and the U.K." ("Report calls for telecom free-for-all in the U.K.," 1981: 49). What advantages? Resale, recall, involves purchase of telecommunications service for resale by a user or specialized company, often with special value-added services attached, and was expressly desired by major users within the U.S. If permitted in Britain, "(i)t would build on advantages within the international cartel by making the U.K. a favorable location from which to conduct business" (Ibid.). For all these reasons, therefore, the British "experiment" is being closely watched by business users, equipment suppliers and other national PTT authorities. (Peltu, 1982: 80–88 *passim*).

Other fissures in PTT control have begun to appear in still other national environments. In Ireland, the PTT has opened itself to private investment—targeted in 1981 to comprise about 100 million pounds, of a total of 220 million budgeted by the PTT ("Irish PTT looks to private sector for telecom financing," 1981: 37–38). Closer to the United States, an interim decision has been made by Canadian authorities permitting interconnection by private systems with Bell Canada, the public network. A separate subsidiary has been established—Bell Communications Systems—to sell new switchboards and terminals to business customers (Block, "1980—a year of major changes," 1981: 66). User group organization continues to strengthen and grow. EUSIDIC is a European association of information services embracing 180 organizations in 27 nations. It provides "a Pan-European meeting ground for representatives of information users, suppliers, telecommunications groups and system operators" (O'Neill 1981: 59), and maintains contact with other user groups. The International Telecommunications Users Group (INTUG), formed in 1974, pushes broadly for "freedom of choice, fair prices" for users (Ibid.: 57). Paul O'Neill explains (Ibid.):

> Until recently, PTT authorities have usually decided what services they will offer and how much they will charge. PTT authorities have told manufacturers what to make and subscribers, who put up the money, have had to put up with equipment, services and tariffs imposed upon them. INTUG members are determined that things will change.

And INTUG, continuing the trend of securing points of leverage within ongoing policymaking bodies, has been accorded observer status within the International Telecommunication Union's Consultative Committee on Telephones and Telegraphs, as of September of 1981. Parallels to INTUG initiatives are evident in the Business and Industry Advisory Committee to the Organization for Economic

Cooperation and Development (OECD). At a High-level Conference On Information, Computer and Communications Policies for the 1980s recently, this business group called upon OECD to support competition over PTT monopoly of telecommunications. The group asserted that "the opportunity to develop, procure, supply and maintain equipment for connection to transmission networks should be available to private business. The flexibility of private business is indispensable in this field" ("Business Calls for End to PTT Monopoly," 1980: 16).

An article on "Challenging CEPT agreements" to allow users greater freedom in their employment of telecommunications facilities within and between EEC member countries, however, concludes that such challenges will not be fully effective "without some political assistance" (Thompson, 1981: 73). Before we can comprehend the nature of the political assistance offered to users by U.S. Government authorities, we must first briefly assess the reasons why political intervention has become so necessary. Why are governments becoming involved in these complicated issues?

"The Politicization of Technology"

As this concerted push by corporate users for "international systems solutions" gathers force, it meets a complex political reaction in more and more nations. Because telematics capabilities will enable vastly greater integration of global corporate activity, foreign companies, whether transnational or not, are threatened. Emergence of telematics first and foremost *within* the United States has given a vital lead to U.S. transnational users, equipment suppliers and, finally, to the U.S. global economic position as a whole. For other countries national telematics policies offer perhaps the only means of influencing the computerization of society to make the process more lucrative and hospitable for *domestic* equipment manufacturers, and *domestic* users whose competitive status is threatened by the rapid assimilation of telematics by U.S. transnationals. Because the scale of economic integration facilitated by telematics is unprecedented—indeed, not yet even predictable—government intervention into telematics by concerted policy development (the Nora/Minc Report in France, the Canadian Clyne Report) is becoming ever more visible and urgent. Such plans and policies are guided by two paramount questions. How is it possible *at once* to deflect some of the momentum of the U.S. transnational offensive in telematics to win back at least a bit more space for constituent economic interests? And, how can the rebuilding of the domestic telematics infrastructure be undertaken so as to ensure a competitive share of burgeoning new markets for the entire range of constituent economic entities, from banks to equipment suppliers? From these basic questions others fan out: How form new allies to ward off IBM? How attempt to respond to politically sensitive labor and employment issues raised by telematics? "Perfectly obvious" to those who discuss communications policy with Western European authorities is the fact "that these questions are now being taken extremely seriously in each of the countries concerned" (Statement Of Dr. Janet Morgan, *International Communications Reorganization Act Of 1981* 1981: 30). U.S. competitors have begun to employ a range of tactics to shape and control telematics; others are pondered. Current economic troubles make such issues all the more important.

For U.S. transnational corporate users, both actual and possible constraints on telematics are increasingly viewed within a larger framework of barriers to trade. "The long term interests of U.S. business and the U.S. economy lie in *free access* and *free entry* into various national economies," recall, claims American Express's Harry L. Freeman (Statement Of Harry L. Freeman, Ibid.: 170, original emphasis):

> To that end, the U.S. must develop a coherent policy and strategy for international information issues. Thus far the U.S. has been reasonably successful . . . or lucky. But luck is no longer sufficient . . . . It is now essential to convince the new Administration and members of Congress that communication policy must be given priority attention.

Re-evaluation of telematics by U.S. authorities has already begun to identify and seek to eliminate a complicated tangle of restrictions imposed by other states on telematics equipment and services and on their use.

Not merely "data protection laws" extending protection of individual privacy (growing in frequency throughout Western nations), but many other, far less tangible constraints are now routinely classed as trade barriers by U.S. authorities. Prohibitions or limits on establishment of U.S. corporate ventures in various foreign markets; special taxes imposed on communications circuits used in specific ways; tariff policies followed by PTTs; procurement preferences by both government and private companies; research and development subsidies for national industries; preferential financing for international procurements; regulatory barriers to introduction of telecommunications and information services industries; technical standards; state promotion of joint ventures to bolster national industry in key areas; export policy—all these have figured in recent U.S. discussions of telematics (E.g., Statement Of Henry Geller, *International Data Flow* 1980: 322–345 especially at 328). The following inventory of trade barriers to telecommunications, data processing and information services, drawn from testimony by the United States Trade Representative's office before a congressional committee, gives a more detailed view of ascribed issues and implications involved in the telematics field.

It must not be disputed, therefore, that what has emerged in this field is a full-blown discussion of international economic policy. Current policy discussions are overlaid by two dominant considerations. First, *any* restrictions upon current transnational corporate activities and, specifically, those relating to centralized and coordinated control of corporate production, marketing, financial and

TABLE 11
Trade Barriers to Telecommunications, Data and Information Services

Major Issue	Country	Motivations for Imposing Barriers	Specific Type of Barrier Applied	Trade Implications
Network controls: restrictions on the availability of leased lines	Japan EC	Increase revenue earnings for PTT's by increasing traffic on usage-sensitive public networks Promote local data processing and database service companies Promote public data communications networks	Various policies and administrative practices to discourage or deny the leasing of private circuits	Increase cost to large users of data communications—present level of data communications may become economically unfeasible In many instances, public network facilities do not provide desired quality of service, thereby reducing efficiency of communications transfer. If private circuits are eliminated, advances in teleprocessing will be severely retarded and many existing teleprocessing systems will be degraded.
Network controls: restrictions on the degree of interconnection of private lines leased by data processing and data base service vendors	Japan This issue remains a potential problem, although Japan, on Dec. 26, 1980 liberalized its regulatory policy in this area.	Increased revenue earnings for PTTS by tying interconnection to use of usage-sensitive public networks Promote local data base and data processing service companies	Private leased lines available to data processing and data service vendors on condition that the circuit only be connected to a single computer system in a single location abroad	Increases cost to users In many instances, public network facilities do not provide desired quality of service, thereby reducing efficiency of communications transfer. Prohibits foreign vendor from marketing full line of services

Network controls: restrictions on the degree of interconnection of private lines leased by TNCS (Transnational Corporations)	Germany	Increase revenue earnings for PTTs by funneling data to usage-sensitive public networks.	Increase cost to users, particularly to smaller users. In many instances, public networks do not provide desired quality of service, thereby reducing efficiency of communications transfer.
		Effective January 1, 1982: (1) International leased lines prohibited from being connected to German public networks unless the connection is made via a computer which carries out at least some processing. (2) International leased lines available only if it is guaranteed that they are not used to transmit unprocessed data to foreign public telecommunications networks.	
Technical/standards barriers imposed on computer and data telecommunications hardware		Promote local manufacturers of computer and data telecommunications equipment Exclude foreign data processing service vendors from accessing local telecommunication networks	Details needed	Affects the ability of companies to export data communications hardware Long waits are common for even the simplest requirements Inhibits quality of data communication system Prohibits or limits opportunity of foreign vendor to provide services

Source: Appendix B to Statement of Geza Feketekuty, Assistant U.S. Trade Representative for Policy Development, Before the Subcommittee on Government Information and Individual Rights, Committee On Government Operations, U.S. House of Representatives, On H.R. 1957, *International Communications Reorganization Act of 1981*, 31 March and 2 April 1981, USGPO: 137–149.

TABLE 11 (continued)

Major Issue	Country	Motivations for Imposing Barriers	Specific Type of Barrier Applied	Trade Implications
Access to Euronet	EC	Promote the information industries of EC member countries	At present, Euronet is pursuing a policy of excluding foreign vendors using computers outside of Europe	Excludes non-EC member country firms from market
Policies which require data processing functions to be performed within a country	Brazil	"Infant Industry" protection for data processing service industries	As part of Brazil's informatics policy, international links for teleprocessing systems are subject to approval by government.	Limits opportunity of foreign vendors to provide services to Brazil
			The principle criteria used in evaluating requests for data links are: (1) Protection of Brazilian labor market (2) Protection of operations of national firms and organizations	Presumably, as Brazil's data processing and software capabilities increase, the Brazilian Gov't will more frequently deny requests for international data links
			All data links approved are reviewed for renewal after three years	Increases cost to users in as much as Brazilian computing services are expensive relative to U.S. vendors
	Mexico	"Infant Industry" protection for data processing service industries	Law in place currently not exercised	Limits opportunity of foreign vendors to provide services to Mexico
			Law giving SCT exclusive rights to provide on-line data processing services	Excludes foreign competition

Policies which require data processing functions to be performed within a country	Germany	Promote local data processing industry	Effective January 1, 1982: Germany will require local processing of data prior to its transmission outside of country.	Restricts business opportunities for foreign D.P. service companies
	Canada	Promote local data processing industry Promote employment of nationals	The Canadian government has recommended to TNCs operating there that data processing of Canadian operations be done in Canada. While only a recommendation, some TNCS have felt constrained to maintain a local data processing facility, although all data processing could be done in the U.S. at a lower price The Canadian Banking Act of 1980 requires banks operating in Canada to maintain and process in Canada all data pertaining to a bank customer The Vancouver Real Estate Board was prohibited from subscribing to a computer prepared "Property Sales" weekly because the data on Canadia real estate provided in the book was processed in a computer outside of Canada	Increase cost to users Restricts business opportunities for foreign data processing service companies

TABLE 11 (continued)

Major Issue	Country	Motivations for Imposing Barriers	Specific Type of Barrier Applied	Trade Implications
Policies which require data processing of financial information to be performed within a country	Germany	Promote local data base and data processing service companies	Bank law has been interpreted by German banking authority as requiring total processing of financial information within the country, thereby excluding the possibility of on-line processing	Increases cost and decreases efficiency of foreign banks
	Korea		Requirement that information about a bank customer be physically retained within the country	Increases cost and decreases efficiency of foreign banks
	Singapore		Requirement that a bank obtain written permission from a customer before it can transmit data about the customer out of the country	Increases cost and decreases efficiency of foreign banks
Restrictive actions related to right of establishment	Brazil Mexico	Develop and protect local industry	Partial local ownership and local content requirements for the manufacture of certain data communications hardware	Eliminates manufacturing and/or marketing of certain equipment; Discriminates against foreign investment
Discriminatory government procurement practices—data processing services	Japan	Promote local industry	Current NTT procurement from Japanese owned companies only	Exclusion of foreign competition

128

		Details needed	
Discriminatory government procurement practices—data processing services			
Import controls on data communications hardware	Brazil	"Infant Industry" protection for computer industry	Restricts foreign manufacturers from market
		As part of Brazil's informatics policy, imports of computers and other data communications hardware are subject to approval by the government. The principle criteria used in evaluating requests for imports are: (1) Protection of operations of national firms and organizations (2) Protection of the Brazilian labor market	Increase cost and operational efficiency of users (example: a bank operating in Brazil was denied permission to import a computer which would have substantially reduced its operational costs. As a result, the bank had no choice but to purchase a domestically available computer with a capacity far in excess of their needs. This resulted in a drastic reduction of anticipated savings.)
Excessive tariff on data communications hardware	EC	Protection of EC member country manufacturers	Discouragement of U.S. exports
		17% tariff imposed by EC on integrated circuits	
		Anticipated efforts to broaden definition of integrated circuits so as to include mounted boards as well	

TABLE 11 (continued)

Major Issue	Country	Motivations for Imposing Barriers	Specific Type of Barrier Applied	Trade Implications
Government support of data processing industry	France, Germany, Italy, U.K., Sweden	Promote local data processing industry	France, 1980–85: $500 million; Germany, 1980–83: $540 million; Italy, 1979–81: $355 million; U.K., 1979–83: $540 million; Sweden, 1979–82: $111 million	Fosters competitive advantage for local firms
Gov't. subsidy of DP industry	(A) Japan	Protection of government provided DCS (NTT)	Absorb operating losses in face of competition	Limited opportunity to compete for other DCS enterprises
	(B) Japan	Foster local industry through government (MITI & NTT) participation in R&D	Joint or sole government ownership or industrial property	Patent licensing by commercial enterprise restricted
EEC (Davignon) plan to promote European-based DP, telecommunications, and microelectronic industries through community-wide standards, procurement and subsidies	ECC	Capture one-third of world market for European-based industry by 1990		Probable discrimination against non-European-based suppliers
Telecommunications rates and rate increases predicated on considerations other than cost	EC, Japan	Avert continued growth of private networks; Promote "in-country" data processing; Promote public data communications networks	Details needed	Increase cost to users; Provide competitive advantage for local data processing service firms

130

Tax on information (value added tax)	U.K.	Promote local information industries Increase revenue earnings	4% duty and 8% VAT (value added tax) levied on all microfilm documents and publications imported from non-EC countries. The levy only affects microfilm brought into the country for resale to the public or to service a client. The tax is applied only to microfilmed information, not its paper equivalent	There is a growing concern that in the future value added taxes will be placed on data processed abroad as a means of promoting local vendors
Restrictions on commercial visas which limit the ability of firms to market and maintain their services	Canada Belgium Brazil Germany Switzerland	Promote local products	Various discriminatory regulations and administrative practices which make it difficult for qualified personnel to enter country	Limits sales opportunities If after sale service is restricted, product becomes less attractive relative to domestic counterpart
Limitation on foreign equity in a service firm	Nigeria Brazil Israel Netherlands South Africa	Promote Nigerian enterprises	Local ownership requirement of 40% for data processing service companies and communications equipment manufacturers Details needed	Discriminates against foreign investment

TABLE 11 (continued)

Major Issue	Country	Motivations for Imposing Barriers	Specific Type of Barrier Applied	Trade Implications
Restrictive foreign exchange regulations	Belgium	Protect local enterprises	Refusal to allow fund transfers or unrealistic limitations on the amount of funds permitted in or out of a country with which to establish or operate data service enterprises	Discriminates against foreign investment
Restrictions on the flow of non personal data	EC	Promote local database and data processing service companies Provide local firms with a competitive advantage over TNC's with regards to local markets	Several European countries, notably the French, are considering data protection laws which would restrict the flow of information pertaining to natural resources, development plans, government operations and budgets, government owned or government supported industries, and certain economic indicators.	Affects the ability of TNCs to obtain the information they need for their foreign operations Restricts sales opportunities of foreign database and data processing service companies
Privacy laws extended to include legal persons	Denmark Austria Norway Luxembourg	Provide local firms with competitive advantage over TNC's		Proprietary business information may be forced to become public information
Misapplication of privacy/data protection laws		Promote local database and data processing service companies		Would greatly impair the business activities of TNCs (particularly information inten-

			sive enterprises: banks, credit card companies, airlines, transportation, securities and money market.) Restrict Sales opportunities of foreign database and data processing companies
Lack of patent and copyright protection	Australia Austria Belgium Brazil Canada France Germany United Kingdom Greece Hong Kong Indonesia Lebanon Israel Saudi Arabia South Africa Philippines Switzerland Sweden Venezuela Mexico Netherlands		Copyright and trademark infringements Details needed

TABLE 11 (continued)

Major Issue	Country	Motivations for Imposing Barriers	Specific Type of Barrier Applied	Trade Implications
Participation in multi-access automated reservations system (S.T.T.) full participation in Air France reservations system	France	Maintain competitive advantage for national airline(s)	S.T.T. system to be available to foreign participation in Air France system. However, French authorities unwilling to force Air France to accept foreign competitors	Loss of market access measured in terms of traffic and revenue
Participation in multi-access automated reservations system (START)	Germany (F.R.)	Maintain competitive advantage for national airline	German authorities take position that matter is outside their jurisdiction	Non-participating airlines are at a severe competitive disadvantage in selling transportation in Germany
Import duties on automated reservations system equipment	Greece	Revenue protect local industry and/or suppliers	31.5% import duty equipment available locally does not meet TWA requirements	High level of duties make it difficult to justify automation
Import duties on automated reservations system equipment	Israel	Revenue protection of local industry and/or suppliers	57% import duty equipment available locally does not meet TWA requirements	High level of duties make it difficult to justify automation
Automated passenger check-in at Rome airport	Italy	Maintain competitive advantage for national airline	Refusal to accept TWA request for automation on grounds that law gives monopoly handling organization exclusive right to render such service	Unfair competitive advantage for Italian airlines

Import duties on automated reservations system equipment	Italy	Revenue protection of local industry and/or suppliers	25% import duty equipment available locally does not meet TWA requirements	High level of duties make it difficult to justify automation
Import duties on automated reservations system equipment	Spain	Revenue protection of local industry and/or suppliers	57% import duty equipment available locally does not meet TWA requirements	High level of duties make it difficult to justify automation
Import duties on automated reservations system equipment	U.A.R. (Egypt)	Revenue protection of local industry and/or suppliers	41% import duty equipment available locally does not meet TWA requirements	High level of duties make it difficult to justify automation
Discrimination in customs valuation between computer and data processing services transmitted through telecommunication network or transferred through physical software products	Brazil Canada United Kingdom Israel South Africa Venezuela		Details needed	
Discriminatory licensing restrictions			Details needed	

USTR Computer Group

investment strategy, *must* be challenged by users. Second, policies which hint at future restraints, or even which create uncertainty about future corporate locational and capital investment plans, must be isolated and rebuffed. "There is major concern," acknowledges the Assistant U.S. Trade Representative, "regarding the potential disruptions and distortions of trade which could occur if governments were to implement proposals currently being considered or if governments were to adopt a restrictive interpretation of laws that have been passed but not yet implemented" (Statement of Geza Feketekuty, *International Communications Reorganization Act Of 1981* 1981: 109–110):

> In many cases such laws provide: a considerable degree of discretion to officials responsible for implementing new regulations, leaving considerable leeway for arbitrary actions by such officials. Beyond these concerns, the uncertainty that has been created by the strong possibility of future restrictive actions by governments, have discouraged many firms contemplating major investments in international communication and data processing facilities from proceeding with such investments.

Precisely because both the character of the emerging telematics infrastructure and the rules governing its future uses are unpredictable matters on which deep dispute separates diverse nations as well as different equipment and services suppliers, firms are hesitant to proceed with investments in telematics (Ibid.: 118–119).

Take the example of Continental Illinois National Bank and Trust Company, as outlined by vice president Robert Walker (Statement Of Robert E. L. Walker, *International Data Flow* 1980: 124–138). After a period of growth through the 1960s and 1970s, Continental is today established in major money market centers with branches or fully-owned subsidiaries. Customers worldwide are supplied with terminals with which they can receive account information, initiate transactions, and engage in many other cash management services such as settlement of intra-company receivables in different currencies, cash flow reporting and foreign exchange and money market rate reporting. Efficient management of the bank's diverse operating units is accomplished by a "centralized data processing center in Chicago" (Ibid.: 126). European branches transmit computerized transaction information daily over a leased line to Continental's communications center in Brussels. From Brussels, data are transmitted at high speed to the central processing unit in Chicago. Each operating unit's records are tabulated by customer and by transaction type, updated, and sent back to branches on a daily basis. Reports thereby created embrace accounting, management information and bank operating data required "to run the banking business at the local level"—out of Chicago! (Ibid.: 126–127). Account officers are provided with data about customers' banking activities worldwide. This enables them "to extract information on a consolidated basis concerning a particular customer, group of customers, country, or region of the world, or any aspect of a group of customers by type of organization, by industry, by size of organization, etc. This information we consider to be of inestimable value in making decisions concerning the proper management of our bank's assets and liabilities and in particular the prudent management of our worldwide loan portfolio'' (Ibid.: 127). Further expansion of the system is under way. But the full value of centralization, Walker stressed, "is only realized when each one of the units throughout the world is included. The inhibition or restriction in any way of data flow from one unit to the head office would degrade the whole concept of this centralized processing system" (Ibid.: 128). Such restrictions, in turn, make the issue of telematics policy and planning of paramount importance to all transnational companies.

Telematics capabilities, as we have seen, will also be required increasingly by foreign transnationals—indeed, probably, by business in general. Upgrading and modernization of telematics networks will therefore assuredly continue. For this very reason, however, vital political issues will also continue to crop up concerning the terms upon which telematics will be implemented within a given national environment. Domestic equipment manufacturers may prefer policies aimed both at immediate economic protection and a future stake in the evolving world economy—as may some interlocked users. Other users may push for a "go-slow" policy out of fear that foreign rivals will otherwise increase their market share of the domestic market. Labor unions, protesting rising unemployment, may urge that similar constraints be placed on the course of telematics innovation—to protect jobs. Throughout, a crucial bargaining chip is available by way of government policies to guide and shape construction of the new telematics infrastructure through technical standards, tariff restrictions, and so on. What John Eger calls "the politicization of technology" by foreign governments in turn compels a political response by U.S. authorities (Statement Of John Eger, *International Data Flow* 1980: 166). It must be remembered, however, that foreign governmental initiatives in the telematics sphere are fundamentally *a reaction* to political and economic decisions affecting the U.S. domestic telematics field. Before turning to some specific initiatives of the Federal Communications Commission, which continues to act as lead agency in the unfolding deregulatory drama, I first inspect the broad contours of an emerging U.S. policy framework for international telematics.

U.S. International Telematics Policy: Coordination and Control

In the face of increasingly deliberate, encompassing attempts by U.S. competitors to shape telematics or informatics or communications plans for controlled national development, the role of the U.S. Government backed by corporate users assumes crucial importance. Four components of this response are apparent. First, the framework for U.S. policy formation is rapidly changing from the relative parochialism of the FCC and a few desks at the Department of State and NTIA to an integrated and well-organized policy function orchestrated by routine intervention by high-level representatives of both the Executive branch and the private sector. Second, a wide-ranging analysis of telematics as an integral component of overall international economic policy has already begun to generate conclusions about appropriate strategy and tactics. Too, as we have seen, a broad and, so far, notably successful, movement is afoot to chip away at public control of national telecommunications infrastructures by government PTT ministries. Although proponents prefer the term "liberalization," this may better be understood as a move for *privatization* of global telematics facilities. Thus, third, concerted efforts by U.S. regulators at the FCC and congressional innovators are aimed at further freeing transnational users to avail themselves of maximally efficient and advanced telematics goods and services. Finally, domestic constraints continue to be loosened as well—not only has AT&T been freed to enter new markets but, too, the "open skies" satellite policy implemented by U.S. regulators in the early 1970s is today rapidly being readied for international application. Above the concerns of any given carrier or equipment manufacturer, these four components share an overarching feature: they are intended first and foremost to serve the expansionary needs of transnational corporate users. Let us turn to the first two now.

"Our present lead in communications technology is not a fact of nature," declared the President's Task Force On Communications Policy in 1968 (*Final Report* 7 December 1968, Chapter I, 5), "but the consequence of history and organization, and the stimulating impact of wartime and postwar government and private programs of research." In the quest for world telecommunications dominance, the report stressed, "(t)he role of government is of unusual importance." "How well government meets its responsibilities is and will continue to be a major factor in the development of the communications system as a whole" (Ibid.: 7).

Continual reassessment and organizational change have been the hallmarks of U.S. Government activity in telecommunications and information fields over recent years. Results have been uneven. Agency responsibilities have frequently come to overlap; confusion over proper areas of jurisdiction has engendered infighting. Yet the various reorganizations—leading to an Office of Telecommunications Policy in the Executive Office of the President in 1970, an Office of Telecommunications in the Department Of Commerce, merger of the two agencies into the National Telecommunications and Information Administration in the Department of Commerce in 1978, and an Office of Plans and Policy at the FCC in 1973, to say nothing of the other telecommunications functions performed by the Departments of Defense and State, and the Interdepartmental Radio Advisory Committee—should not be taken merely as a sign of Government ineffectuality. Telematics policymaking structures have been in flux because the fields they would serve and regulate are undergoing metamorphosis. Overall, Government reorganization attempts to meet the need of transnational economic interests for top-level policy initiatives in telematics—to match the broad Government initiatives faced by these users in other countries, where telematics is under central Government guidance. Coherent policies in the telematics sphere are difficult to achieve, however, because of the "unique position in society" held by telecommunications technology—a position that, by virtue of its scale, interrelatedness, pervasiveness and rapid change "necessarily overrides all the traditional boundaries of governmental authority" (Subcommittee On Communications, Committee On Interstate And Foreign Commerce, *Interim Report And Recommended Courses Of Action Resulting From The Hearings On Telecommunications Research And Policy Development,* U. S. House Of Representatives, 94th Congress 1st Session December 1975, USGPO: 15).

By the mid-1970s, it was being suggested that U.S. world leadership in telecommunications was slipping and that "the Federal Government was partially responsible for this because outdated rules and regulations frustrate technological development as well as competition and therefore limit innovation" (Ibid.: 11). An instruction was given by the Chairman of the House Communications Subcommittee to both the Office of Telecommunications and the Office of Telecommunications Policy to assess and clarify this charge through further investigation. An ensuing report cited regulatory delays as contributing to lack of progress in telecommunications services ranging from direct satellite communications to electronic funds transfers, automated offices and network information services. The report recommended "a massive effort on the part of government and private industry in an attempt to overcome the obstacles limiting technological growth in telecommunications" (Ibid.: 11–12). Available technology should be more intensively exploited "by accelerating telecommunications applications into the marketplace via government, and government-industry actions." Regulatory decisions must enable "earlier diffusion of innovative technology" (Ibid.: 12). Speedy introduction of telematics technology was desirable, even urgent, so much so that the Office of Telecommunications Policy hastened to add that "important policy concerns, on such matters as individual privacy and deregulation for example," should not be "subordinated to a single-minded devotion to market applications" (Ibid.).

To expedite matters the Acting Director of the OTP was asked to convene a Telecommunications Technology Research Advisory Council in December of 1975. The group's purpose was "to assess the government's policy and regulatory process and its ability to keep pace with advances in telecommunications technology" (Ibid.). Such oversight functions are today commonly associated with advisory committees; the number of advisory bodies had grown to an impressive 820 by 1979 (Herman 1981: 215). What is of marked interest about advisory committees—aside from the fact that their deliberations are generally closed to the public—is their composition. The OTP body summoned representatives of Bell Labs, GTE Labs, MIT Labs, IBM, General Electric and Comsat; as Herman asserts of the more general character of advisory groups, they may be not only an effective way for business "in getting its views across and gaining a desired access to government," but also a convenient vehicle "through which businesspeople of the same trade can get together on a regular basis, with the public excluded" (Ibid.: 216). A few of the advisory groupings active in telematics technology and policy fields are: the Industry Sector Advisory Committee on Communication Equipment and Non-Consumer Electronic Equipment for Multilateral Trade Negotiations; the United States National Committee for the International Radio Consultative Committee of the ITU: the United States National Committee for the International Consultative Committee on Telephones and Telegraphs of the ITU: the Space Systems and Technology Advisory Committee; the Space and Terrestrial Applications Advisory Committee; the Federal Information Processing Standards Coordinating and Advisory Committee; the Advisory Committee for Information Science and Technology; the Frequency Management Advisory Council. These advisory groups report to various Government departments and agencies; they tend either to be comprised wholly of private sector personnel, or a mix of private representatives and persons drawn from the relevant departments and agencies.*

If a network of private sector advisory committees today underlies telematics policy formation, the latter has also been assisted by increasing top-level Government participation in telematics issues. The perceived need to streamline and centralize and upgrade Government coordination of telematics has, indeed, greatly intensified. At the close of the Second Session of the 96th Congress, draft legislation was introduced by Congressman Richard Preyer (H.R. 8443) "to reorganize the international communications activities of the Federal Government." Preyer went down to defeat in the election of 1980, but his successor as Chairman of the Subcommittee on Government Information and Individual Rights of the House Committee On Government Operations, Glenn English, introduced a closely similar measure into the new Congress.

The current organization of the U.S. Government, English told the House, "is based on a dangerously obsolete compartmentalization of Government policymaking." The Executive Branch, English charged, was crippled by such divisions and thus was unable to respond to fast-paced international changes in the telematics sector. A yearlong investigation conducted by his Subcommittee had amply revealed "the fragmented, uncoordinated, confused, and conflicting structure within the executive branch for the development and implementation of international telecommunications and information policies" (English in *Congressional Record—House,* 19 February 1981: H532). At stake were billions of dollars and tens of thousands of jobs (Ibid.). Beyond even this, "the health and growth of other economic sectors dependent on the 'information industries' " were equally threatened. Government simply was "unprepared," and lacked both

---

*I should like to acknowledge the assistance of Pamela Schoenwaldt in gathering information on the structure and function of such advisory committees.

"comprehensive plans" and "a coherent strategy for responding to the policies of other nations which may damage U.S. interests" (Ibid.).

The International Communications Reorganization Act Of 1981—H.R. 1957—deserves serious scrutiny not so much for what it will *do*—the bill quite possibly may never be passed into law—but for what it *suggests* about the contours of current thinking concerning the proper organization of Government activities in international telematics. It is suggestive, not definitive; but that which the bill suggests is highly revealing of present policy initiatives. The bill establishes as an explicit state of normalcy an unprecedentedly intimate tie between public and private sectors, permissive of routine intervention in telecommunications and information policymaking by the very highest levels of Executive authority.

A Council of International Communications and Information would be created within the Executive Office of the President. The Council would encompass an Executive Secretary, appointed by the President; the Secretaries of State and of Commerce; the FCC Chairman; the U.S. Trade Representative; the Director of the Office of Management and Budget; the Assistant to the President for National Security Affairs. Many current functions and powers of both the Department of State and of NTIA and the International Communications Agency would be transferred to the Council. The specific mandate of the Council would be to:

> (1) coordinate the policies and activities of all Federal agencies involving international communications and information; and
> (2) review all policy determinations of Federal agencies, and all proposed statements of United States policy by such agencies, relating to international communications and information, and approve, disapprove, or modify any such policy determination or proposed statement.

The Council would be granted extensive oversight powers over Federal agencies, even to the extent of approving, modifying or rejecting policy statements, consultations and policy implementations by Federal agencies (the FCC is expressly exempted from this provision).

Two committees would be convened to advise and guide the high-level Council. An Advisory Committee on International Communications and Information would "provide overall policy guidance." The Advisory Committee would embrace representatives of labor, manufacturers of telecommunications and data processing goods, other manufacturers, providers of telecommunications and data processing services, financial institutions, other services industries, small business, consumer interests, the legal profession and the general public. Rather special powers are conferred on this private sector advisory committee; under Section 5 (c) H. R. 1957 ensures that

> The Council shall, before approving under this Act any statement of new United States policy relating to international communications and

information, consult with the Committee for the purpose of obtaining the views of the Committee on the effect of the proposed submission on the social and economic interests of the United States.

Although not bound to accept the advice or recommendations of the Committee, under Section 5 (e), the Council must inform the Committee of "failures to accept such advice or recommendations," and must also rationalize its conduct before appropriate House and Senate committees. It would therefore appear that the bill expressly subordinates public decisionmakers to private controls. The composition of the private sector advisory committee, moreover, is characterized with unusual precision to include participation by several sectors whose interests in telematics have been most pronounced—banking, services industries and equipment suppliers.

The second committee would subject the Council to oversight by a top-level Government group. This interagency committee, as stipulated in Section 6 (a), would be composed of the Chairman of the Federal Trade Commission, the Secretaries of Defense, Labor and Treasury, the Postmaster General, the Chairman of the Federal Reserve Board, the Administrator of NASA and the Director of the International Communications Agency. This interagency group would be bound to consider and advise the Council (6 (b)) "with respect to problems which Federal agencies encounter in the performance of their responsibilities with respect to international communications and information." Under Section 7, finally, information submitted to any of these new groups—the Council, the Advisory Committee and the Interagency Committee—that is exempted by current law relating to trade secrets and confidential commercial information, may remain largely secret.

H.R. 1957 may never be passed by Congress. But it may not have been designed to do so. Extensive hearings on the bill supply ample evidence that it would provoke further jurisdictional disputes between affected agencies (Kircher, "State Fights to Keep Hold on Communications," 1981: 13). More important, *many of its provisions are already now being taken care of* by informal means. Currently charged with many policy functions in telematics internationally, the State Department, for example, is assisted by a policy level interagency group composed of representatives of the Departments of Defense and Commerce, the Office of the U.S. Trade Representative, the ICA, the Office of Management and Budget, other "appropriate elements of the White House and, depending on the issues, other elements of the Executive Branch" (Statement Of James L. Buckley, Before The Subcommittee On Government Information And Individual Rights, Committee On Government Operations, U.S. House Of Representatives, *International Communications Reorganization Act Of 1981,* 97th Congress 1st Session, 31 March and 2 April 1981, USGPO:299). Intent on preserving his Department's administrative controls, Buckley continued (Ibid.): "I expect NASA, Department of Transportation and the Board for International Broadcasting will be active members on issues relevant to their activities. The FCC participates as an active observer. The group will, in effect, form part of the broader system that is being established in areas of importance to national security." This "senior level policy coordinating working group," it should be observed, is a tangible product of 1980 congressional hearings which first targeted the still-too-fragmented nature of Government procedures for policymaking in the telematics field (Ibid.: 302). The Assistant U.S. Trade Representative, Geza Feketekuty, agreed, "the hearings held last year by this committee had a very positive effect on the degree of cooperation and coordination among government agencies" (Ibid.: 112). And, he assured the English Subcommittee, a private sector advisory system was already "in place which assures private sector guidance throughout the policymaking process" (Ibid.). In this regard, Feketekuty mentioned the favorable impact of a Services Policy Advisory Committee and two Industry Sector Advisory Committees in high technology and services trade (Ibid.). Undersecretary Buckley himself also emphasized that the private sector was already "heavily involved" in various ways with information policy deliberations (Ibid.: 300) and that it would continue to play a formative part in future discussions (Ibid.: 302):

> We will also need the advice of the private sector on a continuing basis. A number of expert advisory groups are already making their collective talent and experience available to the department in particular problem areas. But we do not have today an advisory group to which we can turn for the broad perspectives that are needed for overall policy formulation, including perspectives on future technological advances and their implications. We are considering creating such a group and would encourage its interaction with the existing, more specialized advisory bodies.
> But there is another aspect of relations with the private sector I wish to stress. We must be aware of both the problems and the interests of the private sector if we are to do our jobs effectively. We cannot create viable foreign policy in a vacuum—without full awareness of the private sector's interests. Advisory groups can help fill this gap, but we will place equal stress on encouraging individual firms and organizations to take the initiative in airing their views with us.

H.R. 1957 therefore has already achieved its main goal; its sponsor, Glenn English, has suggested that he will be satisfied "if the bill does nothing more than prod the Reagan administration into making sure the State Department assigns these international issues higher priority" (Kirchner, "International Communications Bill Draws Fire," 1981: 5). Regardless of the fortunes of the bill, the vital issues it addresses have assuredly been made ripe for continuing appraisal by top brass within public and private sectors alike.

There is one lacuna in H. R. 1957—one necessary facet of reorganization to meet the challenges of international telematics policy that it does not display. This unmet need was articulated by William Colby in testimony before the same Government Operations Subcommittee (*International Data Flow* 1980: 160): "We need to organize the supporting staffs," declared former CIA Director Colby, to "form alliances among potential beneficiaries in academia, the information industry and its customers . . ."

Scarcely a year later—on 8 April 1981—Congressman George R. Brown of California introduced legislation to rectify this shortcoming. H. R. 3137, the "Information Science And Technology Act Of 1981," is again more suggestive of current thinking than of hard-and-fast realities. The bill aims to set up an organization within the Executive branch to "provide a focus and mechanism for planning and coordinating Federal research and development activities related to information science and technology" (Kirchner, "Hearings Open On Bill Forming Info Institute; NTIA Negative," 1981: 13). This proposed Institute for Information Policy and Research would try to overcome current fragmentation of Federal research, development and policy activities by bringing together effectively public and private interests "to discuss national information concerns in a cooperative forum" (Section 2 (8)).

H.R. 3137 is opposed by agencies on whose jurisdiction it would trespass—notably, NTIA and the National Science Foundation. In light of continuing budget cuts across nonmilitary Government departments, its chance of passage seems slight. Yet the bill's striking attempt to cultivate "private and public mechanisms to coordinate the introduction of information and telecommunications technologies into factories, offices and schools" must be noted in any case. The Institute would provide both to the Government and to private sector affiliates "data and information not otherwise available about developments and trends in information science and technology throughout the world, including the efforts of foreign governments to develop and articulate national policies" (Title I, Section 104 (1)). The Institute would, moreover, assist in acclimating the population of the United States to introduction of advanced information technology and services—for the necessity of rapid introduction of the latter into U.S. society is never questioned. It is stipulated that the new center shall "conduct and support research into the broad policy issues concerning human interaction with, and acceptance of, information technology in the home, school, and workplace" (Title I, Section 104 (3) and (4)). The Institute itself is directed (Ibid. (11)) "to serve, to the extent practicable, as a model for the use of information technology, by exemplifying in its organizations and function the employment of this technology to enhance efficiency and to promote personal satisfaction and self-fulfillment." Such enhancements are to be overseen by a National Information Science and Technology Board, whose fifteen members "shall at all times include representatives of private sector businesses providing information products or services or trade associations comprised of such businesses," as well as members of scientific and educational and professional groupings (Title I, Section 105 (c)). Whatever its legislative fate, the proposal is another sign that efforts are well advanced to develop that "most important" process, through which "we must begin to pull ourselves together domestically to build consensus on the importance of the general issue at hand and on priorities and strategy for own future activity"—as the Office of the U.S. Trade Representative puts it (Draft, "A Preliminary Review of Barriers to Trade in Telecommunications, Data Processing and Information Services and Transborder Data Flows Motivations for Imposing Barriers, Trade Implications and Possible Approaches to Resolution," 3 September 1980: 25–26). Such initiatives, however, once again merely ratify an already widely prevalent practice. As Dale Hatfield of NTIA, admittedly trying to preserve his agency's jurisdiction in the face of the bill's mandated changes, told the congressional committee in hearings on H. R. 3137, "virtually all of the desirable objectives and shared goals of this proposed legislation can be achieved . . . and, indeed, are being pursued within the framework of existing laws and organizations" (Kirchner, "Hearings Open On Bill Forming Info Institute; NTIA Negative," 1981: 13).

Organizational restructuring within Government has been paralleled by newly emergent *strategic* initiatives. At the base of current strategic planning is intransigent insistence upon free trade—a policy stressing maximum freedom for corporations to invest and disinvest globally. Deputy Treasury Secretary R.T. McNamar, for example, emphasizes that the Reagan administration—like every postwar government before it—"is committed to pursue a policy of open markets" and expects the same from U.S. trading partners. "If they don't but instead resort to a variety of protectionist devices, impair our access to their markets or artificially undercut the competitive advantage of U.S. industries, we won't stand idly by," McNamar told a group of institutional investors ("U.S. Trading Partners Are Warned on Use of Barriers and Subsidized Export Loans," 1981: 6).

Protectionism is on the rise; newspaper reports of mounting international trade disputes have found steadily more emphatic headlines set in progressively larger type. "U.S., Common Market on Blowup Path," signals a recent *Wall Street Journal* article (Pine, 1981: 35)—and then goes on to tell of a "confrontation"—in agriculture, textiles and steel—"that neither side wants," but that neither "knows how to avert." Faced with recession, European governments confront strong protectionist pressures from home industries, while the Reagan administration, "bent on widening markets for American companies, is pressing aggressively to loosen current trade restrictions."

Protectionism is most especially disturbing when it affects what Assistant Trade Representative Feketekuty labeled "our most competitive service industries"—"communications, data processing, and information services" (*Service Industries Development Act* S. 3003, 96th Congress 2d Session, 24,25 September 1980: 10), for not only are these industries a top-growth area in their own right, they also comprise a strategic infrastructure for virtually all transnational business. In the current era foreign governments, according to Feketekuty, are "uncertain about the full scope and public policy implications of the new technology in telecommunications and data processing, and the steps they may have to take in the future to protect various public policy goals" (Appendix A to Statement of

Geza Feketekuty, Before The Subcommittee On Government Information And Individual Rights Of The Committee On Government Operations, 31 March and 2 April 1981, Hearings on H. R. 1957, *International Communications Reorganization Act Of 1981* USGPO: 130). For this reason the "key policy dilemma" is simply that "any push at this stage to negotiate new international rules to eliminate uncertainty about future government actions"—a critical requisite for corporate planning—"would likely result in international agreements that are fairly restrictive, reflecting the uncertainty governments face regarding future policy requirements" (Ibid.: 131). Premature efforts today, that is, "could result in a more restrictive final outcome" (Ibid.). On the other hand, "the natural evolution of events in the absence of any international understandings could lead to a highly restrictive trend, and would do nothing to reduce the uncertainty faced by enterprises contemplating new investments" (Ibid.). Thus, restated, the policy challenge will be "to devise an approach to international negotiations that will minimize the risk of a restrictive trend and reduce the uncertainty faced by business, while not forcing governments to commit themselves in great detail on policy issues that remain uncertain, pending further development of the new technology and the evolution of public opinion" (Ibid.).

Technology develops most felicitously in an environment wherein it is thought to be largely value-free. Such an atmosphere must more and more be created and nurtured; it no longer merely appears as the natural child of a nineteenth century intellectual legacy of belief in progress through science. As Feketekuty emphasized, thus (Ibid.: 132), "an ideological debate over the relative virtues of different national philosophies and policy approaches is less likely to result in mutually beneficial, commercially-oriented accommodations." Technology and technological applications should be simply accepted—not debated—despite their inherent dependence on socially patterned uses and structures of organization and control. The first objective of U.S. policy in the telematics field therefore must be to deflect discussion as far as possible into technical rather than substantive political circles.

Beyond this general objective U.S., strategy is at best only tentative. Nonetheless, there are some indications of prospective policy directions. A preliminary agenda for implementation in the delicate telematics field, for example, has already been articulated by the U.S. Assistant Trade Representative. The agenda is of interest, once again, for what it reveals about the underlying character of policy concerns and not insofar as it represents a solid and fixed plan.

First, owing to the potential problems associated with "discriminatory implementation of domestic regulations," some sort of regulatory standards code might profitably be considered (Ibid.: 133). Such a code would furnish "some basic principles and a set of domestic and international procedures" for dispute settlement (Ibid.: 134). It might be aimed at overcoming discriminatory use of privacy regulations, data protection rules, standards governing establishment and operation of telecommunications systems, and restraints placed on connection of data processing and telecommunication equipment to public telephone systems (Ibid.). The code thus would establish "an obligation to seek the least distortive design and implementation of domestic standards" (Ibid.).

Second, since "one of the basic issues" concerns "the right to plug other equipment into the public telecommunications network," an internationally agreed " 'right to plug in' equipment interface that meets agreed technical standards" should be negotiated" (Ibid.: 134). This "right" would embrace not only the ability to attach equipment meeting agreed standards but also the ability "to sell services that can be provided using such equipment" (Ibid.).

Third, since international telecommunications traffic rates may be set by PTTs so high as to discourage use of foreign data processing or data base services "or in other ways to discourage the use of international services," through discriminatory taxation and the like, corrective action is desirable (Ibid.: 135):

> While each government should have the right to decide how prices for public services are established, or what taxes are imposed, it might be possible to agree that prices should reflect costs, and that the establishment of user charges or taxes should not be used to erect barriers to trade in services.

Finally, because "a number of countries are moving towards outright restrictions on the use of foreign data processing or data base services," negotiating approaches need to be devised "for limiting and removing such restrictions" (Ibid.).

The essence of this policy initiative is identical to both the regulatory programme developed and largely adopted within the United States market and the core of the international business user agenda. Cost-based pricing of telecommunications, the right to offer new services reliant on specialized interconnect equipment and, above all, the ability to centralize global services so as to take advantage of resources located within one nation when developing markets elsewhere—these comprise the necessary precursor to further expansion of user-oriented networks and services.

Effective U.S. Government intervention in pursuit of these goals has become increasingly linked to "reciprocity"—an eye-for-an-eye strategy stressing that as U.S. companies fare in foreign markets, so shall representative firms from those same markets fare within the United States. We should retain existing regulatory powers to assure the maximum degree of reciprocity in our overseas telecommunications relationships," Robert Bruce recently asserted (*International Communications Reorganization Act Of 1981*: 61). Formerly of the FCC, Bruce argued that that agency should be given "discretion to disapprove the entry into U.S. markets of foreign owned or controlled service providers" (Ibid.: 58). (We shall find below that the Commission relies upon its existing mandate to force concessions on foreign correspondents in the telecommunications facilities planning field.) Indeed Bruce urged Congress to provide "an explicit grant of authority to deal with reciprocity issues broadly—i.e., both refusals to allow entry in foreign markets by U.S. service providers and power to deal with entry of foreign entities in U.S. markets" (Ibid.). Such coordinated development of policy, he insisted (Ibid.), "would convey to our telecommunications partners a firm resolve to establish an open, flexible, and mutually acceptable framework for the provision of international telecommunications and information services." In S. 898, the Telecommunications Competition and Deregulation Act Of 1981, passed by the Senate in October, 1981, it is explicitly mandated that

> For the purpose of ensuring fair and equitable treatment of United States telecommunications enterprises seeking access to foreign markets, the Federal Communications Commission shall have authority to conduct inquiries and establish policies, rules, regulations, and requirements applicable to the entry of foreign telecommunications carriers or foreign entities supplying information services into domestic United States telecommunications markets upon terms and conditions which are reciprocal . . . .
> (Title II Section 240, as printed in *Congressional Record—Senate*, 5 October 1981: S 11030)

The reciprocity concept has immediate application as well to the important field of telematics equipment procurements by government ministries. Government control over procurements is a key weapon in global telematics competition insofar as the rebuilding process is concerned, for expenditures of this class are enormous. Nippon Telegraph and Telephone Public Corporation, for example, the Japanese national telephone company, buys about $3 billion worth of equipment each year ("Nippon Telegraph Is Buying Abroad," 1981: 27). Often, procurement policies aim to bolster domestic equipment suppliers by contracting for required instrumentation and plant from home companies even where lower bids are received from foreign competitors. However, this form of protectionism is of uneven scope; in 1979, for example, not less than 82 percent of the $3.1 billion European Economic Community public sector market for computer equipment was held by United States companies (Malik, 1980: 25–26)—while within the United States, U.S. firms controlled 98 percent of the public sector computer market. United States companies' shares of many overseas markets is already so great that legislative action to enlarge their opportunities overseas could easily set off a trade war, with dire results ("The Reciprocity Boomerang" 1982: 30).

Nonetheless the U.S. Government has been quite active in seeking out further opportunities for domestic telematics equipment interests. In December 1980, agreements reached during the Tokyo round of the multilateral trade negotiations for GATT—the General Agreements on Trade and Tariffs—were finalized after U.S. negotiators successfully compelled their Japanese counterparts to open Nippon Telegraph to foreign suppliers. It may be noted that this at least nominal victory for U.S. equipment companies was won only over the opposition of Nippon itself, whose president resigned over the issue (Van Zandt, 1981: 26–27). As a private company, of course, AT&T is not presently bound to accept a similar measure of Government intervention—in fact, "AT&T is notoriously loath to buy telecommunications equipment not developed by Bell Laboratories or made by Western Electric, its manufacturing arm" (Chace, 1981: 1). It presumably needed little pressure from U.S. Government authorities, therefore, when in the wake of intensifying protectionist impulses internationally, AT&T recently withheld a contract for fiber optic communications equipment from Fujitsu, the Japanese telematics manufacturer that had outbid seven other competitors, and, citing national interest, gave the $75 million contract to its own Western Electric unit ("Japan Intends to Go on the Offensive, Ask U.S. To Drop Its Own Barriers to Trade," 1981: 33).

Reciprocity is a bargaining tactic used to further *specific* policies, normally favorable to users and, in important degree, also to equipment and service suppliers. Beyond the general framework implied by the U.S. Assistant Trade Representative, above, development of U.S. international telematics returns concretely and continually to the deliberations of the Federal Communications Commission and Congress.

THE UNITED STATES REGULATORY AND POLICY OFFENSIVE

At the very same time that policy initiatives were taken to throw open the U.S. market to advanced telematics equipment and services, similar moves began to be made in the sphere of international policy. This was far from coincidental; the vast domestic market is now, first and foremost, merely a vital and influential zone within a functioning *world* economy unified and coordinated by transnational corporate decisionmaking. Yet because of its size, conditions of corporate performance within the U.S. have frequent and sometimes far-ranging impact on the global economy—"spillover effects."

It is in this context that the Federal Communications Commission has moved consistently over the past fifteen years or so to cultivate more favorable *inter*national circumstances for heavy users of telematics goods and services. Policies have been aimed, like their counterparts in the domestic realm, not negatively at restraining the carriers— even if this is their sometime result—but positively, at ensuring maximum efficiencies and economies for transnational users. Broadly speaking, the overall results of changing Commission policies have been few but profound: elimination of an accustomed series of market divisions aimed largely at preserving oligopoly in provision of international telecommunications services, whose major effect had been to impede innovation and implementation of advanced telematics services; and integration of disparate categories of service to permit greater economies and efficiencies to large users. Again like the domestic policy scene, these results may be mistaken by some as engendering competition. Although competition is frequently a byproduct, however, the policies are not changed in pursuit of competition per se—but rather, only in pursuit of the advanced, integrated, economical services which transnational corporate users demand. Here I merely sketch the outlines of some of the more significant of these initiatives. An attempt to integrate voice and record—or non-voice—services, has been the FCC's first such move.

Integrating Voice and Record Service

The dichotomy between voice and record services springs from the earliest history of the contest between AT&T and Western Union, the former with patents to the telephone, the latter with rights to the telegraph. Growing rivalry between the two companies led to an agreement in 1879, "leaving telephones to Bell and telegraphy to Western Union, effecting a duopoly in the telecommunications business and formalizing the separation of modes" (Grad and Goldfarb, 1976: 410). The agreement proved fortuitous for AT&T; less so for Western Union.

A second entrenched division (of importance because it is intricately intertwined with the first) between domestic and international record service, followed from the eventual near-failure of Western Union. In 1943 Congress permitted the latter to merge with its rival, Postal Telegraph, also near bankruptcy. Pursuant to creation of this new domestic monopoly telegraph company Congress fashioned three provisions designed to prevent it from unfairly competing against the overseas record carriers (Ibid.: 415–417). Congress, in Section 222, an amendment to the 1934 Communications Act, ordained that Western Union should divest itself of its *inter*national record operations to assume the mantle of monopoly at home. Twenty years later, Western Union complied, and spun off its Western Union Cables subsidiary; the latter took the name Western Union International (WUI), and today carries no corporate tie to its original parent. (WUI has just been acquired from Xerox by MCI.) Section 222 also mandated that, in concert with the FCC, the international record carriers (IRCs) should determine a formula for distribution by themselves of messages destined for foreign points originating with Western Union. Finally, the statute established five *gateway cities*, to designate the points at which Western Union might interconnect with the international record carriers. Any larger number of gateways would have as its immediate result a decrease in message traffic carried by Western Union—as its domestic monopoly would have been breached by the IRCs. Three major IRCs were, accordingly, permitted to pick up and return record signals at New York, Washington, D.C., and San Francisco (ITT, RCA and Western Union Cables). A subsidiary of United Fruit, Tropical Radio Telegraph (now TRT Telecommunications), was given gateway rights to New Orleans and Miami—best suited to its extensive interests in Central America.

At first, international record carriers were divided into those utilizing submarine telegraph cables and those employing radio to transmit signals; initially, too, various IRCs provided direct services only to limited and differing areas of the world. Both of these distinctions were breaking down by the 1940s, as IRCs made use of *both* cables and radio for back-up assistance in the event of cable failure, and as mergers and acquisitions expanded the global scope of the resulting few IRCs (*International Telecommunications Policies*, Hearings Before The Subcommittee On Communications Of The Committee On Commerce, Science, And Transportation United States Senate, 95th Congress 1st Session, 13 July 1977, Serial No. 95-54, USGPO 1978: 132).

With its greater bandwidth requirements, transatlantic telephony was conducted until 1956 by over-the-air radio transmission. Radiotelephony afforded indifferent service owing to a proclivity for signals to fade in and out with changing weather conditions, making conversation difficult. While record communications were divided between Western Union in the domestic United States (defined to include Mexico and Canada) and the IRCs, voice telephony was the ken of AT&T operating in tandem with the independent telephone companies inside the U.S., and alone in the overseas market.

Despite the unpredictability of overseas radiotelephony, use of the system skyrocketed. Intensified use nurtured concern to improve techniques utilizable for international telephony; as an AT&T spokesman says, "(t)he public—and the Defense Department—wanted a higher quality communications medium" (Nichols, 1980: 361). The key technical obstruction to development of voice-grade submarine cables was the lack of a reliable repeater, used to prevent degradation of signal strength over long distances. AT&T solved this problem and, in 1956, laid the first transatlantic cable (TAT 1) adequate for telephonic communication with cooperation from British and Canadian authorities.

With emergence of voice-grade undersea cables, the technical division between voice and record international services was effectively eroded. Owing to their enlarged bandwidth, submarine cables could carry telegraph, telex, facsimile and data traffic, and at increased speeds, as well as voice messages. A shift in U.S. military telegraph leases from the IRCs to AT&T's new cable TAT 1, prompted the IRCs to appeal the question of who might offer record services to the FCC. In what is known as its TAT 4 decision of 1964, the FCC resolved the question (AT&T Co., 37 FCC 1151). AT&T might continue to build voice-grade cables, the Commission decided, but it must offer ownership shares in them to the IRCs based on their prospective *use*. Although voice and record traffic would be routed through the same cable circuits, however, AT&T was prohibited from furnishing record services—which would return to the IRCs. The result of the TAT 4 decision was to sustain a technically arbitrary dichotomy between voice and record services for all users except the U.S. Department of Defense—which was permitted to obtain integrated voice and record service circuits *only* from the IRCs. For other users, separate terminals and facilities were required to transmit and receive voice and non-voice signals—resulting in added expense.

In the meantime, AT&T was eventually allowed to offer a striking new service in the domestic market—Dataphone service. Dataphone service permits subscribers to ordinary telephone service to send data over the telephone network at regular rates. It is highly useful to those desiring intermittent or occasional nonvoice service. Once the domestic Dataphone service was undertaken, in the FCC's words, there was "increasing public interest in the availability of overseas dataphone service." The purpose of its Overseas Dataphone inquiry launched in 1972, therefore, was "to determine whether dataphone service should be authorized to foreign countries generally and, if so, whether the telephone or the record carriers, or both, should be permitted to provide the trans-oceanic facilities for this traffic" (*FCC Annual Report* Year Ended 31 December 1973, USGPO: 128).

It was a familiar alliance between equipment manufacturers and large corporate users which got things started and pushed toward an ultimately favorable—for them—decision. A letter from Xerox Corporation initially requested the FCC to "consider removal of the present limitations upon the international transmission of data and facsimile signals" (Comments Of Xerox Corporation Pursuant To Notice Of Inquiry, FCC Docket No. 19558 "Inquiry into Policy to be Followed in Future Authorization of Overseas Dataphone Service," 29 September 1972: 1 (original letter, 17 June 1971)). Xerox began by conceding that its interest in expanding the user-public's capacity to use data services was "parochial," in that the company marketed facsimile and data communications equipment on several continents (Ibid.). Yet it was vital in any case, declared Xerox, that the FCC "should strive to give telephone users as many options for use of the public network as are technically and economically feasible" (Ibid.:4). The details of this suggestion are of some interest, owing to the relationship outlined between domestic and international telecommunications in its behalf.

Users, announced Xerox (Ibid.), presently could employ the telephone network on a domestic basis for "freely interchangeable voice and non-voice transmission." "We believe," the company continued, "that they should be permitted to conduct international transmission under the same rules which apply to domestic transmission" (Ibid.). Present tariff restrictions preventing users from utilizing telephone circuits to transmit data and facsimile signals should be considered "a significant and unwarranted impediment upon users" (Ibid.) with quite negative consequences. "The result of this restriction is to require American users having need for international data or facsimile transmission or reception to either lease dedicated international private lines or rely upon significantly more cumbersome procedures utilizing multiple carrier facilities" (Ibid.):

> Many users cannot economically justify a dedicated international private line. They do, however, have significant requirements for demand type service. In addition, many users have requirements for international service from multiple locations within this country which cannot be met by the more limited point-to-point service of a private line. Such service would be available if interconnection with the public telephone network were permitted.

Geographically dispersed enterprises with numerous facilities *within* the U.S. Thus were being kept from obtaining maximally efficient telecommunications service between their various domestic facilities and corresponding overseas locations. "Currently available procedures using multiple carrier facilities for international transmission of data and facsimile signals," Xerox went on (Ibid.: 4-5), "are likewise unsatisfactory to meet the needs of many users.

> Those procedures presently involve a recording and retransmission of the signal at each international gateway and result in such loss of time, flexibility and message quality and confidentiality as to be unacceptable even without consideration of the costs involved.

The ensconced divisions between voice and record, and domestic and international services were too unwieldy. Xerox recommended that the FCC pursue a strategy that was "not 'either/or' in nature." Permission should be granted for *both* international transmission of facsimile and data over the public switched telephone network *and* direct interconnection of record carrier facilities with the telephone network "as an additional option for users who have a need for international data services" (Ibid.: 5). In brief, only provision of both types of service "will effect optimum customer benefit."

Xerox had the foresight to append letters it had solicited from several corporate users on these matters (Ibid.: Appendix). Allied Chemical Corporation, for instance, wrote to assert that, while the company's use of overseas dataphone was at present not urgent, "we can see a future requirement for this ability" (Ibid.: Letter dated 8 September 1972). Allied preferred that its prospective need be met "by use of the voice network and interconnection of the international record carriers" with AT&T's public switched network (Ibid.).

Eaton Corporation (Ibid.: Letter dated 7 September 1972) responded that there was "a definite and growing need for international transmission of facsimile and data within this company." Indeed, the company declared (Ibid.), "Eaton has or will have computer centers in the United Kingdom, Germany and Italy that should be linked to the Telecomputer Center here in Cleveland." Eaton again came down in favor of redundancy, arguing for "the combination of both voice network and the interconnection of the international record carriers" for these transmissions. Eaton indicated, as did other corporate users polled by Xerox, that it had no objection "to Xerox constructively using this information as part of its presentation to the FCC."

Boeing Computer Services (Ibid.: Letter of 6 September 1972) could conceive of no valid technical reason "why international facilities should be treated differently from our domestic or U.S./Canada facilities." Boeing peremptorily explained (Ibid.):

> Certainly none of us that depend heavily on facsimile for our day-to-day domestic record communication should be reluctant to state rather categorically

that in the future this same convenience should be available in our international operations. In a similar vein, those of us engaged in computer service operations can see no legitimate technical reason for not extending our "dial-up" teleprocessing services beyond our national borders. In a small degree, at least, our planning for international data transmission has already been limited by the current restrictions, since initial low volume usage is most economically handled on dial-up facilities prior to a commitment to dedicated lines.

As we shall find again and again, the dual advance of technical capabilities and user demand within the domestic telecommunications industry acted as an opening wedge in arguments for expanding analogous services into international markets. (Use of the domestic market as an international economic bludgeon has a long history in communications; Shayon (1977) has chronicled it in the late 1920s and 1930s for radio equipment manufacturers.)

It should also be underscored that loosening of previous restrictions mandating separation of voice and record carriage was justified by recourse to the public interest clause in the Communications Act. As the Electronic Industries Association declared (Comments of EIA, FCC, Docket No. 19558, "Inquiry into Policy to be Followed in Future Authorization of Overseas Dataphone Service, 29 September 1972: 2–3), "the American public has become accustomed to dataphone-type usage as a part of his telephone call capability. There is no technical impediment to dataphone-type usage on overseas calls, and the American public should not longer be denied this logical extension of conventional telephone usage." The public interest would not be served, warned EIA, should only the IRCs be permitted to offer such service (Ibid.: 2). The Computer and Business Equipment Manufacturers Association urged that (Ibid.: Comments of CBEMA, 29 November 1972, 3) "the ultimate framework fashioned by the Commission be one in which users are granted the greatest flexibility possible in their choice and potential uses of the services authorized." Agreeing, the Central Committee On Communication Facilities of The American Petroleum Institute told the FCC (Ibid.: Comments of API, 29 November 1972, 4–5) "to promptly resolve those regulatory issues requiring disposition for the provision of competitive and compatible facilities for overseas dataphone . . ."

The Commission eventually determined that AT&T should be permitted to offer dataphone service internationally, and that the IRCs might expand their existing dataphone-type services and interconnect them with AT&T's domestic network (*FCC Annual Report* Year Ended December 31 1977: 81; 57 FCC 2d 705 (1976)). The user-equipment manufacturer alliance received everything for which it had hoped. As FCC Chairman Richard E. Wiley announced, the Commission had concluded that "the best interests of the consumer rest with the availability of flexible, efficient and cost-effective record communications using the international telephone network alternately for both voice and non-voice communications." Such use, he continued, would be privately beneficial without being publicly detrimental "and would be consistent with the Commission's longheld view that the public's use of the public network should be made as flexible as possible" (*International Telecommunications Policies* 1978: 109).

Like other policy changes, the Overseas Dataphone decision attempted to redress what the Commission called "outmoded international telecommunications policies, time-consuming agency procedures, and long-standing industry practices (which) were impeding the availability of international communications services and increasing the costs borne by the American consumer" (Ibid.: 106). Its effects on those key industry participants, the carriers, were clearly subordinated to these larger goals. During hearings associated with a continuing attempt to rewrite the 1934 Communications Act, it was protested by the international record carriers that erosion of the voice-record dichotomy will betoken new competition between affected carriers. Allowing AT&T to enter the market for international record services, in the eyes of RCA Global Communications president Eugene F. Murphy, "would open the door to AT&T's total domination of the international communications market" (Statement Of Eugene F. Murphy, *The Communications Act Of 1979* Vol. 1 Part 1, Hearings Before The Subcommittee On Communications Of The Committee On Interstate And Foreign Commerce, U.S. House Of Representatives, 96th Congress 1st Session On H.R.3333 Titles I and III General Provisions, 24,25,26 April; 1,2,3,4,8, May 1979, Serial No. 96-121, USGPO 1980: 333). Telephone service revenues, Murphy protested, already comprised two-thirds of total international service revenues. Congress should take care not to permit Bell to compete for the remaining one-third, because its domestic near-monopoly granted it enormous discretionary market power internationally. Murphy asserted (Ibid.: 336) that Bell might use its vast domestic network as a competitive club against the IRCs—for, he hinted, the latter were having difficulty obtaining interconnection on an equivalent basis with AT&T operating firms.

In a separate proceeding the FCC currently is soliciting comments on whether AT&T should be allowed to furnish international record services generally—and whether the IRCs in turn should be authorized to provide international voice carriage. Such a result of course would altogether abolish the industry structure that for decades relied upon the voice-record dichotomy to preserve customary perquisites. And, in a Notice of Proposed Rulemaking in Docket No. 80-632 the FCC has concluded tentatively that the policy enumerated in the 1964 TAT 4 decision barring AT&T from providing record services internationally should be set aside. These moves show clearly that the Commission is now willing to cultivate a new industry structure—with fierce competition between established carriers, perhaps, and with the prospect of further industry consolidation that true competition usually confers—in order to furnish transnational users with the services they desire. The same result is obvious in light of analogous erosion, as a consequence of deliberate policy, of another primary industry division—between domestic and international telecommunications service.

Integrating Domestic and International Service

Let us briefly review the merger legislation of 1943, whose several provisions—intended to prohibit Western Union as a new domestic telegraph monopoly from employing its massive domestic base to favor its own overseas telegraph company, Western Union Cables, over other IRCs—set the course of the international record industry into the 1970s (Statement Of Richard E. Wiley, *International Telecommunications Policies*, 1978: 133).

First of all, Western Union was required to divest its international telegraph operations with "due diligence"; full divestiture took place in 1963. Pending divestiture, Congress secondly mandated that the Western Union Company should distribute outbound unrouted international message telegraph traffic among all the overseas carriers in accordance with a "just, reasonable, and equitable formula in the public interest" (Ibid.). This provision was intended to deter Western Union from diverting unrouted traffic—messages for which the sender does not designate a specific carrier—to its own cable facilities in preference to those of other IRCs. It basically tried to freeze each carrier's market share at its pre-merger level by fixing a quota of traffic for each carrier comprised of a mix of routed and unrouted traffic. A carrier whose share of routed traffic increased was to give up an equivalent amount of unrouted traffic—so its total share would be broadly stable. Not unexpectedly the formula system "discouraged the IRCs from aggressively competing for routed traffic" (Ibid.: 134) and, instead, cultivated the prevailing oligopolistic market structure.

Third Congress distinguished between Western Union's legitimate service area and that of the IRCs by drawing a line between domestic and international service and by designating domestic points or gateways where the IRCs might receive messages for international delivery and where Western Union could pick up incoming traffic from the IRCs. The line was curiously drawn. "Domestic" service circumscribed the North American continent including Canada and Mexico. "International" service similarly embraced anomalies: Hawaii, Puerto Rico and the Virgin Islands were designated international points, although these territories or protectorates bore closer formal ties to the U.S. than did Canada or Mexico. The original gateway cities, as has been mentioned, were five—New York, San Francisco, Washington D.C., Miami and New Orleans.

These basic policies endured into the 1970s despite the fact that, in Wiley's words (Ibid.: 140), "numerous developments have rendered many of those wartime decisions anachronistic." What had changed?

Western Union International had finally entered the picture as an independent firm in 1963. More vital, message telegraph service was suffering a pervasive decline in the face of explosive growth by telex and private line services. The trend continues. Between 1970 and 1979 overseas telegraph revenues collected by the IRCs dropped from $53 million to $38 million, while revenues for popular business services like telex and private line increased from $63 million to $299 million, and from $49 million to $96 million, respectively ("Revenues From Overseas Communications Services Reported by Telegraph Carriers," Table 25, *Statistics Of Communications Common Carriers*, Federal Communications Commission Year Ended 31 December 1979, USGPO: 158). Telex and private line are preferred by large users—transnational corporations and governments—because both allow greater flexibility, efficiency and speed than does ordinary telegraph service. The formula intended to preserve the relative proportions of telegram traffic enjoyed by each IRC in 1942 thus became no longer relevant to the newer meat-and-potatoes services they provided.

Although, as we have seen, originally each IRC carved out a part of the globe as its more-or-less exclusive domain, by the mid-1970s RCA, ITT and WUI had direct circuits to an increasingly broad series of communication centers (TRT retained a more restricted service area) (Statement of Richard E. Wiley *International Telecommunications Policies 1978*: 142). Each IRC moreover sought to furnish users end-to-end service—this was a major incentive for the expansion just noted—and accordingly supplied telex terminals and associated equipment and supplies on a "bundled" basis. Bundling of prices means that a set of potentially separable services, equipment and supplies are packaged together under a single undifferentiated price tag. This effectively locked users into acceptance of complete service from a given IRC—for telex terminals for instance, could not possibly be obtained from independent suppliers at competitive rates. A not-so-incidental side effect of bundling practices is that it becomes practically impossible to find out how much each element or component of service individually costs.

From a large user's point of view, therefore, a number of pressure points existed in international record services. Rates were set at an uncertain level; users knew only that telex was very lucrative for the carriers, but they had no idea of how much each portion of this bundled service contributed. Geographic dispersal of major users both within and outside the United States, meanwhile, added additional pressure because as matters stood only a few international gateway cities were authorized to route and return messages from domestic to international points. Western Union took in substantial revenue for routing record messages to and from the international gateways; should the number of gateways be expanded, however, users might send messages much shorter distances before being picked up by an IRC—and pay less as well. (Western Union claimed in 1980 that "better than 50% of its telex business consists of traffic transmitted to and from foreign points" (Hirsch, 1980: 36). Richard E. Wiley summed up the situation faced by users in congressional testimony (Statement Of Richard E. Wiley *International Telecommunications Policies,* 1978: 110):

Since no single carrier serves all overseas locations, in order to reach all overseas points a customer must subscribe to telex service from at least two different IRCs and use a separate telex machine for each carrier. The consumer's use of telex service is severely restricted by the lack of interconnection among the IRCs' networks and the inability to use a single standard telex terminal. Furthermore, although a customer has the right to buy a telex machine, he or she is discouraged from doing so by the IRCs' bundled rates which require only a very minimal charge for use of the telex machine.

The Commission, declared Wiley, was "exploring the possibility of requiring that the various IRC and Western Union telex networks be interconnected and that separate compensatory charges for the telex terminals and transmission be instituted" (Ibid.). His rationale for this attack on oligopoly in international record service was that (Ibid.) "(i)n this manner unnecessary costs and impediments to consumers will be identified and, where appropriate, removed."

There resulted a scramble by all industry participants to maximize the benefits of reorganization—and to cut losses when possible. In 1971, RCA filed an application for 15 new gateway cities; other applications by other IRCs followed, until a total of 21 new gateways had been requested. The Commission declared tentatively that conditions in the industry had greatly changed since 1943, and that the carriers no longer limited their services to a discrete overseas service area. Nor did technical needs any longer dictate location of transmission facilities only at U.S. coastal cities. The FCC therefore proposed to permit all IRCs to operate in all five existing gateways immediately (58 FCC 2d 250 (1976)). The Commission simultaneously stated that in a separate inquiry it would consider whether the carriers should be authorized to operate in any additional cities beyond the original five (54 FCC 2d 532 (1975)). Such expansion in turn raised the question of economic impact upon Western Union, which would also be taken up in the proceeding.

Not to be outdone, Western Union too attempted to expand its service offerings. Shrewdly utilizing as a test case a point designated by Section 222 as international and hence outside its area of service, but which had become a domestic point in all other legal respects, Western Union filed to extend its domestic Mailgram service to Hawaii. The Commission ruled in favor of Western Union in 1975 (55 FCC 2d 668); the IRCs appealed its decision; and on 20 October 1976, it was reversed by the Court of Appeals on the grounds that Section 222 permanently barred Western Union from any form of international telegraph operations—and that because under its terms Hawaii is an international point, the company was ineligible to furnish Mailgram service to Hawaii (544 FCC 2d 87; Statement Of Richard E. Wiley, Ibid.: 160). Litigation thus developed between Western Union and the IRCs over the terms upon which expansion and integration of the sundered markets would occur.

The dynamic force behind the IRCs' requests for additional gateways, however, and indeed, behind much of the subsequent shake-up in international record service, was the development of satellite technology to meet large corporate and governmental users' integrated service needs. Satellites permit sophisticated data services—not only telex, but computer communications and facsimile—over wideband digital circuits. Domestic satellites, recall, had been authorized early in the 1970s and were making rapid progress as a preferred medium for such integrated traffic. As matters stood, however, *domestic satellite transmission from (or nearby) company premises within the U.S. had to be routed to one of five gateway cities and thence to an international satellite uplink over conventional analog circuits.* The tremendous efficiency gained by satellite carriage was interrupted and severely impaired at this one juncture, where a domestic satellite signal had to be transferred to inferior analog circuits for only a few miles before being retransmitted over international satellite facilities. The plight of one company that spelled out its position to the FCC may make this point clearly.

Counsel for Texas Instruments wrote the Commission to state that his client "urges the Commission to act promptly to add Dallas and Houston, or Etam, West Virginia, to the list of international gateways for private line services" (Letter From John Bartlett, 23 November 1979, FCC Docket No. 19660, "In the Matter Of International Record Carriers' Scope Of Operations In The Continental United States,": 1). TI's corporate headquarters are convenient to Dallas and Houston. Were the latter cities to be authorized as international gateways, technically advanced satellite services might be utilized by the company on an end-to-end basis. Alternatively, Etam, West Virginia, houses an earth station for international satellite traffic. As TI explained, the company required "high speed data circuits between its headquarters in Dallas and points throughout the world" (Ibid.). Under current "outmoded gateway limitations," however, it was forced to route messages through the Washington, D.C. gateway via ITT-operated limited analog circuits to Etam, where the international satellite uplink allowed ITT to transmit its signals on to Europe. The weak link was the short hop from Washington, D.C., to Etam—and it created intolerable service deficiencies given the potential of advanced satellite transmission techniques. As Texas Instruments insisted, "(d)elay in expansion of the gateways can only result in continuation of unsatisfactory service to the public." (Ibid.).

Three weeks after this missive, the FCC adopted what then-Chairman Charles D. Ferris labeled "more pro-consumer policies in the field of international communications than we have since the FCC came into existence" (FCC Orders FCC 79-840 through 79-847, Separate Statement Of Charles D. Ferris, Chairman, "RE: Federal Communications Commission's Polities (sic) In The International Telecommunications Area," 75 FCC 2d 382). The Commission allowed the IRCs to expand service within the U.S. through authorization of 21 additional gateways *and* permitted Western Union to move into international communications service (Ibid.: 383). And the FCC now gave users the option of connecting with IRCs directly, rather than through AT&T or Western Union, if such connection was cheaper or of better quality (Ibid.: 382–383). Most of these decisions entered the Courts for resolution.

Strikingly, however, the FCC placed the interests of domestic and international record carriers behind those of major users. In exchange for allowing the IRCs to serve directly 21 additional cities, and for permitting Western Union into international services, the Commission resolved that the carriers must interconnect their domestic access facilities with each other and with Western Union. The IRCs, too, would be forced to unbundle their tariffs—henceforth, they must charge separately for transmission, access lines and terminals. Aside from prompting a hailstorm of litigation over the terms upon which these provisions would occur, the decisions proffered decisive support to users.

> The basic aim in both cases was to reduce costs for the user. Interconnection would give the user access, through a single terminal, to points served by one IRC but not by others, thus eliminating the need to subscribe to more than one service and lease more than one terminal as previously required.
>
> Unbundling would enable the user to exploit discounted rates introduced last year by all of the IRCs. These tariffs allow a discount of 55 cent/min if the customer provides his own terminal and access circuits to the nearest IRC termination point. (Hirsch, 1980: 36).

The FCC had unilaterally introduced conditions favoring freedom of choice for major users—who now, according to Ferris, might "comparison shop for different parts of international service—terminals, access to international switches, and international transmission" (75 FCC 2d 382). Further parallels with newly achieved U.S. domestic policies are again remarkable.

Cost-based pricing has been elected to provide the best marker for international charges (FCC Docket No. 80-339, "In The Matter Of Revisions To Tariffs For Establishing Separate Charges For Terminals, Tielines, And Transmission Offered In Connection With International Telex Service And Implementing Expanded Gateways And Additional Domestic Operating Areas For International Telecommunications Service," *Memorandum Opinion And Order* Released 8 August, 1980).This move follows rates-of-return investigations by the FCC that determined that pretax profits in 1976 ranged from 32% to 58% for international telex (59 FCC 2d 240 (1976); Statement Of Robert E. Lee, Acting Chairman, FCC, *International Record Carrier Competition Act Of 1981*, Hearings Before The Subcommittee On Communications Of The Committee On Commerce, Science, And Transportation U.S. Senate 97th Congress 1st Session On S. 271 "To Repeal Section 222 Of The Communications Act Of 1934," 18 February 1981, Serial No. 97-5, USGPO 1981: 8). Early signs are that prices for the largest users are indeed falling in the wake of these decisions. *Business Week* reports that "users of the approximately 150,000 teletypewriters now installed in the U.S. should benefit from competition in telex services. Prices for international telex transmission have fallen 20% since July (1980—DS) because of a price war that followed the FCC's action" ("Deregulation roils the telex market," 1980: 66). Again, we find real price competition within an oligopolistic industry—as a direct result of Government actions on behalf of large users.

Equally vital, the longstanding division between domestic and international record service is now on the verge of extinction. "Many customers need both domestic and international data communications service and would prefer to meet all their needs with a single source," asserts one knowledgeable industry participant (Statement Of Philip M. Walker, GTE Telenet, *Hearings On The International Record Carrier Competition Act of 1981*, 1981: 37). A bill to repeal Section 222 has passed both the Senate, and the House, and was in 1982 signed into law by President Reagan. FCC and court decisions have been supplanted with a firm new mandate for integrated domestic and international data service.

It may be noted, too, that trends toward cost-based pricing and unbundling of service at once affect correspondent relationships between U.S. carriers and the PTTs with which they must collaborate. New competition at this end in effect inflames a similar trend at the other. Integrated data communications services at the cheapest possible prices are the desired goals—and the entire international telecommunications industry is being (sometimes unwillingly) transformed to obtain them. Current attempts in this field, however, are not limited to a merely negative function; for AT&T is being groomed to take on a thoroughgoing transnational role.

AT&T International

Nowhere is the attempt to restructure international communications more important, and apparently neglected, than in the case of AT&T. While Western Union and the international record carriers quarrel among themselves, all of their bickerings pale in comparison with questions raised by the entry of AT&T into provision of merged telematics services in the international arena.

Ranked by sales, AT&T is the fith largest company in the United States as of 1980; ranked by profits or by assets, it is the first ("The Forbes 500's," 1981: 260). *Yet the AT&T goliath is not among the top one hundred U.S. transnational corporations*; the second biggest domestic telecommunications firm, GTE, in contrast, was in 1980 the fiftieth largest U.S. multinational company ("The 100 Largest U.S. Multinationals," 1981: 92–94). GTE has attained this formidable transnational position although it rules only about 8 percent of the U.S. domestic telecommunications market. Western Electric, the Bell System's manufacturing arm, which is the twenty-second largest industrial corporation in the United

States, is one of but two of the largest 25 industrials that does not also appear among the top 100 U.S. multinationals ("The 500," 1981: 324). AT&T must be treated as highly anamalous within the corporate structure of the United States: it has been conspicuously absent from transnational activities. The Bell System has achieved its legendary profits—$6.9 billion in 1981—almost entirely within U.S. domestic markets (although this of course must also include AT&T's monopoly share of overseas telephone traffic). *Current regulatory, legislative and executive initiatives are aimed, however at freeing Bell to become the fully transnational entity that its mammoth size would necessitate, were it to be engaged in virtually any other line of business.* Opening up the domestic U.S. market to increased competition—essential for users' needs—has of late thus been followed and paralleled by attempts to free the major domestic company, AT&T, to enter various international markets, to permit transnational corporate users to take advantage of its immense market power in an international arena. Users, for example, may and do avail themselves of IBM's wares in global markets—and achieve numerous economies, presumably, by centralizing their purchases in this manner. However, they have been prevented from obtaining ready access to the other telematics leader, AT&T. Users' demands for global telematics "solutions" are now finding use as arguments to permit the Bell System to become a fully integrated transnational corporation. A second chief reason to push AT&T into global competition stems from the elephantine size of the international telematics equipment market. As the rebuilding process outlined above continues, Arthur D. Little estimates that between 1980 and 1990 as much money will be spent on telephone equipment alone as was spent between the invention of the telephone and 1980—a period of 100 years ("ITT: Groping For A New Strategy," 1980: 77). For United States equipment and services firms not to maintain and, if possible, extend their market share, in this booming field is not short of suicidal.

Again because of the dynamic nature of telematics, the scope of the information business is without precedent. Neither Bell nor its domestic and international competitors any longer perceives hard and fast barriers between previously disparate varieties of equipment and service—and the race is on to achieve a degree of integration in information industry activities that, even a decade ago, would have been unthinkable. "More and more it is apparent," an AT&T spokesman told a House panel recently, "that there are no functional boundaries" within what AT&T now terms "the knowledge business" (William W. Betteridge, Statement Before The Subcommittee On Telecommunications, Committee On Energy And Commerce, U.S. House Of Representatives, 97th Congress 1st Session 28 May 1981 *Status Of Competition And Deregulation In The Telecommunications Industry*.: 566). "(T)here really is no longer a "domestic market" separated from international dealings" (Ibid.). Therefore, any attempt to segment or bracket portions of the integrated global market "is to place artificial and inefficient limitations on the users of its products and services" (Ibid.: 565).

AT&T is today intent on penetrating the global business communications market—encompassing "the totality of a customer's information management needs" (Ibid.: Statement Of William P. Stritzler, AT&T, 460). The size of this market is immense—in AT&T's estimate, $660 billion—though it may be noted that some inflation of this figure would serve Bell's purpose by diminishing its own already prodigious domestic role within this now-much-larger field. It includes voice telephony ($190 billion); face-to-face meetings ($280 billion); document communications (mail, filing, etc., $170 billion), and about $20 billion for data communications, image communications and video communications (not counting video broadcasting) (Ibid.: 461). The information business is thus enormous, reflecting the technical integration permitted by telematics; indeed, still according to AT&T, business expenditures on the above-mentioned categories of the information market account for between 20 percent and 80 percent of business operating expenses, depending on the kind of business (Ibid.).

One may quarrel with AT&T's figures. It is not possible, however, to contest the assumption behind them. This is that information business vendors and suppliers are today compelled, as a result of technical integration fueled by large user need, to view the six modes—voice, data, image, video, face-to-face and document—as a single, manipulable field of commercial endeavor. "The fundamental issue that the information systems user must manage is the way in which he selects among modes of communicating and the extent to which he deploys technology to enhance those modes" (Ibid.: 462). Better information management will translate into "big dividends for users" (Ibid.: 464). Thus it is not alone owing to the global reconstruction process but also because of the unprecedented integration of information management within corporate structures that "a successful vendor must have the ability to help the customer optimize among the use of all his communications modes" (Ibid.: 462).

AT&T's vision gives a clear indication that the company plans ultimately to compete actively across the entire range of the information business. "The tools available to manage information in this market include tools of transport, switching, processing, entry, storage and retrieval applied in each of the three phases of communication: Preparation, Movement and Reception. It is vital for a vendor in the Information Industry to be permitted to make full use of all of these tools" (Ibid.: 462-463). Today, declared AT&T's Assistant Vice President of its Market Management Division, the business customer sees installed on his premises (Ibid.: 465):

1) a telephone system and "wires",
2) heating, ventilating, air conditioning and energy system and "wires",
3) a data processing system,
4) data communications "wires",
5) links to service bureau time sharing systems,

6) word processing equipment,
7) experimental electronic mail systems,
8) conferencing systems (perhaps video),
9) provisions for accessing a voice and data network.

By and large each one of these separate systems '' requires different boxes, different power sources, different transmission methods in many cases, and different methods and procedures generating substantial training burdens'' (Ibid.). In the future, however, the business customer ''sees each of these systems coming under his control through communications links and software programs with uniform terminal interfaces that are easy to learn and use.'' The dynamic need *''to integrate these functions through increasing substitution of electronic for manual modes is the essence of the customers' emerging need to manage information''* (Ibid., my emphasis). It is this breathtaking integrative impetus to which vendors must be responsive (Ibid.: 465).

(It should be mentioned that in AT&T's view ''(t)his same scenario applies in varying degrees to smaller businesses and even to residences'' (Ibid.). The home will be the ultimate—though today still only nascent—site of information industry services; according to another AT&T representative, this home of the future will contain ''a total integrated Home Management System consisting of three main integrated subsystems''—environmental management, which is directed at home security and energy consumption; resource management, aimed at retail shopping, banking and other commercial and financial services; and information management, which is centered on information/entertainment options (Ibid.: Statement of Dennis J. Sullivan, Jr., 342). It is due to this convergence of energy, banking and finance, and information-entertainment sectors upon the home that tremendous conflicts are now surfacing between various prospective service suppliers—newspapers, broadcasters, cable system operators, computer services companies, banks, retailers, and so forth.)

To sell to this epic market the Bell System has been gearing up for entry into the international telematics field. As early as 1975 it was recounted in a congressional report that the single largest U.S. telecommunications equipment manufacturer, Western Electric, had ''shown recent interest in pursuing selected international market opportunities''—although their nature was not elaborated *(Interim Report And Recommended Courses Of Action Resulting From The Hearings On Telematics Research And Policy Development,* Subcommittee On Communications Of The Committee On Interstate And Foreign Commerce, U.S. House Of Representatives, 94th Congress 1st Session, December, 1975, USGPO: 13).

In the wake of the FCC's Second Computer Inquiry AT&T undertook what may well be one of the largest corporate reorganizations in business history. The portion of the prospective change accorded most scrutiny was the formation of a new separate subsidiary, ''Baby Bell,'' to compete in enhanced services and customer premise equipment. The parallel change that has received virtually no play in the trade press concerns formation of a second new subsidiary, itself probably to be located *within* "Baby Bell." This second subsidiary is AT&T International, "which will seek new 'global' markets for Bell System products and services" (American Telegraph & Telephone Company, *Annual Report* 1980: 21).

According to Robert E. Sageman, president of the new firm, over the last fifty years "Bell's international efforts were sporadic at best. Most of our resources and energies were directed at the regulated domestic market.

> During the last few years, though, the whole of the telecommunications industry has changed so dramatically that we had to reconsider many of our axioms—because they weren't axioms any more. We asked ourselves why we should restrict our business to just this country. We know that, for years, companies such as IBM, GTE, and Ford Motors have been generating a major portion of their earnings abroad. So, we asked: Why should we limit ourselves primarily to the United States? (Chase, 1981: 11).

What motive force did Sageman single out as vital to this evolving new perspective? "(I)t's apparent that the multinational corporation is an increasingly important economic factor and we should be able to serve that kind of corporation. Business is an international phenomenon: We should be out there in the international marketplace" (Ibid.). At present AT&T International's activity is apparently concentrated in Saudi Arabia, South Korea and Taiwan, but planning is mainly "complete" to enter a wide variety of foreign markets (Ibid.: 12). A commitment exists at Bell Labs and Western Electric "to produce products that will truly be world-class entities" (Ibid.: 13):

> Any product developed by the Bell System will be more profitable when we can market it all over the world. In fact, some products may be profitable *only* because there are overseas markets to supplement domestic earnings. So AT&T International may be the critical factor in some of the go/no-go decisions.

The "transnationalization" of the world's largest corporation thus bids fair to become a reality soon—and one that is not lost on international competitors.

Current policy gives every promise of being directed at placing AT&T at the competitive edge of global information markets. "The need for AT&T . . . to enter overseas markets became apparent the last 10 years as the information industry grew worldwide," declared even the normally staid and subdued *1981 U.S. Industrial Outlook* (Department of Commerce 1981: 476). The recently settled antitrust case against the Bell System, however, furnishes the best example of how the push to develop a transnational AT&T has affected domestic telematics policy.

Along with military security, a major argument employed by AT&T to prevent break-up of its vertically integrated business concerned "the impact that divestiture would have upon the productivity of the economy at large and the position of American industry in world markets.

> Quite simply, the time is past when the preeminence of this country's industry in virtually every world market can be assumed. To the contrary, the United States has lost its leadership in many important areas of worldwide economic activity—steel, automobiles, transportation, and electronics are but a few examples . . . .
> Telecommunications is one of the few remaining examples of extraordinary productivity at home and world leadership abroad. At a time when many industries are pressing the Government for protection against foreign competition, the Bell System not only holds its own against Siemens of Germany, Erikson of Sweden, Nippon of Japan, and other foreign industrial powers, but continues to be the leading force in technological innovation in the entire world . . . . It simply makes no sense to dismantle the structure that has made the Bell System the envy of the world, and then substitute in its place a restructured industry whose members would likely end up pleading for protection against the very foreign competition that Bell now trounces.
> (Civil Action No. 74-1698 In U.S. District Court for District of Columbia, USA, Plaintiff, v. American Telephone And Telegraph Company; Western Electric Company, Inc. and Bell Telephone Laboratories, Inc., Defendants, Defendants Pretrial Brief, 10 December 1980: 208–209).

AT&T's tremendous research, equipment manufacturing and telematics service capabilities, far from being dismembered, should be consolidated to permit Bell's speedy entry into foreign markets. It was highly laudable, announced a *Wall Street Journal* editorial after the antitrust settlement ("Ma Bell Unbound," 1982: 32), that AT&T would now be free to bring its most sophisticated technologies to market, wherever "opportunities arise."

Although within the United States the international ramifications of the settlement have been downplayed, they have not been lost on analysts overseas. The outcome of both the AT&T and the IBM cases, declared the British journal *The Economist* ("Stripping Ma Bell," 1982: 14), clearly showed "that in antitrust enforcement America will in future examine the world market, and American self-interest in it, rather than concentrate myopically on domestic concerns."

In the wake of the settlement, AT&T itself has moved swifly to confirm the accuracy of this judgment. Scarcely two weeks after resolution, AT&T brought up the possibility of turning to foreign capital markets for future debt issues (in 1982 the company expects to raise at least $4.5 billion in debt and equity markets). AT&T also announced it "may build manufacturing facilities abroad, conclude marketing agreements and increase its overseas presence by purchasing shares in more foreign concerns" ("AT&T Says It May Look Abroad to Offer Some Debt and Build Production Plants," 1982: 8). AT&T is clearly moving rapidly to accept the transnational stature for which it has been groomed.

Not only the mere fact of settlement, however, but also its particular *form*, points to the growing hold over domestic telematics policy exerted by global concerns. A "new international perspective," suggests *The Economist* ("Stripping Ma Bell," 1982: 14), "may have influenced the trustbusters to break off from AT&T the local telephone companies rather than Western Electric," the powerful manufacturing arm of the Bell System. The Justice Department, recall, sought not merely divestiture of the Bell operating companies, but actual dismemberment of the vertically integrated system itself. "Not only does the Government seek the separation of Western Electric and the 'allied research and development facilities' of Bell Laboratories from the Bell operating companies," protested AT&T at one point in the case, "it also seeks the separation of the Bell operating companies from Long Lines" (Civil Action No. 74-1698, In The U.S. District Court For The District Of Columbia, United States Of America, *Plaintiff*, v. American Telephone and Telegraph Company; Western Electric Company, Inc.; and Bell Telephone Laboratories, Inc., *Defendants. Defendants' Third Statement Of Contentions And Proof* 10 March 1980: III, 2098).

"(P)reserving Bell Laboratories and Western Electric as integral parts of AT&T," the modified Consent Decree, explained AT&T Chairman Charles L. Brown, " . . . promises to be a key element in maintaining America's world leadership in the Information Age and in the technology on which that leadership is based" (Statement of Charles L. Brown, U.S. Senate Committee On Commerce, Science, And Transportation, 97th Congress 1st Session, 25 January 1982: 3). However, the settlement also places at risk the technical and economic well-being of the operating companies, and may also help to engender staggering local rate increases, all "to assure in this era of worldwide competition that the benefits of the Information Age come first" to transnational business, if not "to America" per se ("AT&T Share Holders Newsletter," 1982: 1).

The growing tie between domestic and international telematics policy is again evinced in a speech by then–FCC Chairman Charles Ferris (Ferris, 1980: 30):

> We are witnessing now grave crises in critical industries like steel and automobiles which a decade ago were leading sectors of our economy. We should not forget that the roots of many of our current industrial problems were not of recent origin. They stem in large measure from our past failure to recognize the perils of protectionism and to anticipate the future needs of our society.
>
> Today telecommunications and information industries are poised to play the same leading roles in our economy that the steel and automobile industries played ten years ago.
>
> I am determined that we not repeat in the communications industries of this country these errors of the past. I am detemined that we not allow an environment that impedes innovation in our information services and allows oth-

ers to surpass this country's leadership in the fields of telecommunications and information handling.

Such sentiments presumably also came to the forefront of the Second Computer Inquiry, where Ferris's FCC determined to permit AT&T entry into provision of unregulated computer communications, through a separate subsidiary, and thereby developed a conceptual template for subsequent policy initiatives.

The Telecommunications Competition And Deregulation Act of 1981—S. 898—for example, includes provisions which, while barring AT&T from offering enhanced services and equipment within the domestic market other than through a separate affiliate company, explicitly permit the firm maximum latitude in marketing such advanced services and equipment overseas:

> a fully separated affiliate may enter into any agreement, contract, joint venture, partnership, or any other manner of business to sell, market, or service exports with any United States entity or entities provided such activities take place outside of the United States . . .
>
> The conduct of business between a dominant-regulated carrier or any affiliate of the dominant-regulated carrier and any fully separated affiliate shall not be based upon, or include, any preference or discrimination arising out of affiliation. The provisions of this subsection shall not be construed to apply to facilities or services which are sold for use, consumption, or resale outside the United States.
>
> The restrictions or requirements on any dominant-regulated carrier or affiliate contained in this section are not applicable to transactions involving commodities or services, which are sold for use, consumption, or resale outside the United States or the research, development, and manufacture of those commodities (S.898 Section 227 (c) and (d) and Section 228 (i), in *Congressional Record—Senate* 5 October 1981: S11026).

Government attempts to make AT&T relinquish its protected domestic monopoly have been matched neatly by efforts to bestow on the behemoth firm an aggressive transnational presence. Implications for world markets—including the domestic U.S. market—of a transnationalized AT&T are of profound significance. At a minimum, they entail emergence of a twin giant to stand beside IBM in global battle. As Anthony Smith has seen (Smith, 1980: 139), of course "(i)n forcing its transnational giants into an intensified competition, America is in effect inducing them to challenge . . . all of the countries across which the American giants will conduct their economic warfare." This would appear a risk worth taking if integrated, international solutions for U.S. business users are the chief desired goal.

The Satellite-Cable Dichotomy and Its Uses

If, partially on behalf of transnational corporate users, AT&T is today being groomed to adopt a fullblown transnational presence itself, parallel activities in the communications satellite field may themselves further destabilize established arrangements in international telecommunications. Policies fashioned in support of satellite systems by United States authorities have created substantial pressure already on PTT ministries overseas to restructure institutions in accord with transnational corporate demands. Brief review of satellite history must precede amplification of this claim.

Developments in rocketry, spacecraft and communications by the late 1950s made it feasible to contemplate construction of a global satellite system (Richard E. Wiley Statement *International Telecommunications Policies, 1978;* Grad and Goldfarb, 1976). Numerous advantages of satellites were apparent to various interested parties. The U.S. Department of Defense saw in them a desirable means of extending and coordinating an increasingly global military power. Electronics, computer and telecommunications equipment manufacturers were delighted at an opportunity to inject their revenue stream with funds derived from satellite procurements and private sales. Common carriers hoped to incorporate satellites into their established systems—on their own terms and at their own pace. Generally, a U.S. satellite system entity was viewed by Government and industry leaders alike as having great potential in spearheading an international satellite development program responsive to U.S., rather than European or Soviet, interests (Schiller 1971: 127–146).

In 1962 Congress chartered a private corporation, the Communications Satellite Corporation, or Comsat, defining it to be a common carrier subject to appropriate provisions of the 1934 Communications Act—and regulable by the Federal Communications Commission. (Oversight of Comsat's relations with foreign governments to assure that these would be consistent with U.S. policies was delegated to the President). Congress further instructed the FCC in regulating Comsat to "engage in such ratemaking procedures as will insure that any economies made possible by a communications satellite system are appropriately reflected in rates for public communication services" (47 U.S.C. Section 721 (c) (5), quoted in Statement Of Richard E. Wiley *International Communications Policies:* 136).

Comsat was called into being to lead development of an international satellite system, and lost no time in doing so. In August of 1964, the International Telecommunications Satellite Organization or Intelsat was formed. Comsat was the official United States signatory to an organization whose membership was, and today remains, dominated by public governmental authorities—PTTs. Yet Comsat, a private company emerging from and operating within a gray area between public and private sectors owing to its unique foreign policy role, according to erstwhile FCC chairman Wiley held effective veto power "over all of INTELSAT'S major decisions" (Ibid.). Comsat itself puts it somewhat more delicately (Communications Satellite Corporation *Annual Report 1980:* 8): "the Corporation became the prime mover behind the creation of . . . the International Telecommunications Satellite Organization."

At the outset the U.S. common carriers were allowed to purchase slightly less than half of Comsat's total stock, the result being that through its formative years the new entity was influenced decisively by the established interests of AT&T, ITT, GTE and RCA (Schiller, 1971: 130). Initial ownership influence by the terrestrial carriers over Comsat was sufficient to allow them to gain benefits in two crucial FCC decisions. In the first, the Commission concluded that the terrestrial transmission facilities employed to carry traffic to and from the various U.S. international gateways and satellite earth stations were *not* properly portions of the earth station complexes, and should therefore be provided by the terrestrial carriers and not by Comsat (2 FCC 2d 658 (1966)). As we found earlier, this created an obstruction to integrated end-to-end international data transmission service by satellite. Second, and possibly more vital, was the FCC's "authorized user" decision. Again bowing to the interests of the terrestrial carriers, the Commission concluded that, although Comsat *might* be authorized to provide satellite service directly to non-common carrier entities, it was to be primarily a carrier's carrier—and hence in ordinary circumstances, users of satellite facilities should be served by the terrestrial carriers. Acting as intermediaries, that is, the terrestrial carriers might lease appropriate satellite circuits from Comsat and only then might the latter be made available to final users (4 FCC 2d 421, (1966): 6 FCC 2d 593 (1967)). Only in unique conditions might Comsat be authorized to provide service directly to non-carrier users. The terrestrial carriers benefited from the authorized user decision, of course, because it effectively ensured that Comsat would be unable "to use its satellite monopoly to compete with existing U.S. carriers for international traffic" (Statement Of Richard E. Wiley *International Telecommunications Policies:* 113). In turn, the terrestrial carriers as intermediaries would be able to influence the rates, technical features and future plans of Comsat. Composite rate averaging of satellite and cable circuits by the terrestrial carriers had dictated that satellite users pay for service at a price reflecting not only the costs of providing satellite service, but also the investment of the terrestrial carriers in cables—even though satellite circuits are unquestionably cheaper.

These arrangements were to the liking of the terrestrial carriers—and within strict limits (to be discussed later), also of their foreign correspondents, the government ministries of Posts, Telephones and Telegraphs. There are obvious incentives for both groups—terrestrial carriers on the U.S. side and PTTs abroad—to prefer cables to satellites, and therefore to try to limit and constrain the latter. The capital invested by a terrestrial carrier in cable circuits, most notably, may be included in its rate base to earn the going rate of return. (As rate-regulated companies, the common carriers earn a profit based on their capital investment, instilling in them a general propensity to over-invest where feasible.) Satellite circuits, necessarily *leased* from Comsat, are treated as expenses upon which no return is authorized. Indeed, the very satellite traffic projections of future demand on which Comsat must base its own investment plans have been supplied by the only entities authorized to lease its facilities—the carriers. The PTTs prefer cable to satellites for equally compelling reasons.

First, as Wiley has asserted, a given country "may have a significant industrial interest in the manufacture of cable, but not in satellites" (Ibid.: 114). Such an interest might be accommodated by the prime cable contractor—AT&T prudently provided for it in some manner in construction of several transatlantic cables. On the other hand, satellites were from the first deliberately intended to break control by European nations—most importantly, by Great Britain—over international communications (Schiller 1971: 136–137) and to supplant it by an American-dominated system. *Fully 90 percent* of Intelsat's expenditures for space segment facilities between 1964 and 1978 have "been spent or committed in the United States" (*Science, Technology, And American Diplomacy 1981* 97th Congress 1st Session, Second Annual Report Submitted to the Congress by the President Pursuant to Section 503(b) of Title V of Public Law 95-426, USGPO, May 1981, Committee On Foreign Affairs, U.S. House of Representatives: 259). As James Hodgson, the British PTT's representative to the European Conference Of Posts and Telecommunications (CEPT)—a Pan-European body active in international facilities planning and policy—stated in a letter to erstwhile FCC Chairman Charles Ferris: "As is well known, the USA has for most practical purposes a monopoly of the provision of the space segment of the Intelsat system" (FCC Docket No. 18875, "In The Matter Of Policy To Be Followed In Future Licensing Of Facilities For Overseas Communications," 25 November 1977). Second, and closely related, while cable facilities were largely "bilateral operations between the U.S. carriers and selected foreign PTTs," (Statement Of Richard E. Wiley *International Telecommunications Policies:* 113), satellite planning is not generally perceived as closely responsive to the interests and concerns of foreign PTTs. Although Comsat's formal control over Intelsat has diminished, its ownership interest having declined from 61 percent in 1964 to its present 23 percent (Communications Satellite Corporation *Annual Report 1980*: 8), this is still not an insubstantial stake. Further, although Comsat's initial role as system manager for Intelsat has been altered by Definitive Agreements so as to allow the Director General of Intelsat to assume formal day-to-day management of the system, Comsat today furnishes research and development and technical and support services to Intelsat on a contractual basis. To quote a recent FCC proceeding, therefore, Comsat's role in Intelsat has shifted from being "pervasive" through the developmental period, to "significant" in the current era (FCC Docket No. 79–266, "In The Matter Of Comsat Study—Implementation Of Section 505 Of The International Maritime Satellite Telecommunications Act," Final Report And Order,

Released 1 May 1980: 6–7). For all these reasons, as Richard Wiley discreetly hinted (Statement Of Richard E. Wiley, *International Telecommunications Policies:* 114), a given foreign correspondent "might prefer the larger proportionate voice in bilateral cable ownership and operation to the more limited and diffuse participation possible in a global satellite consortium."

A curious alliance has therefore often marked the international communications field. The U.S. terrestrial carriers—ATT, ITT, RCA, WUI, TRT—and their PTT foreign correspondents alike, and for deepseated economic and strategic reasons, prefer cable to satellite circuits—in spite of the economies and efficiencies to users afforded by the latter. As Wiley puts it (Ibid.: 114), "it cannot simply be assumed that U.S. carriers would inevitably choose those international facilities which result in the lowest overall cost to the American ratepayer—particularly when the varying interests and involvements of their foreign correspondents are factored in."

It was during deliberations over prospective need for a new transatlantic cable—TAT 5—beginning in 1968, that the potential conflict between Comsat and major users on one side, and the terrestrial carriers and their PTT correspondents, on the other, boiled over for the first time. The Commission set about weighing operational and cost advantages of satellites and cables, respectively, in order to decide which—or which combination—would best serve the public interest. In the process, the Commission imposed a novel procedure on the U.S. terrestrial carriers, demanding that they refrain from signing any final agreement with their overseas partners. The latter, of course, being sovereign authorities, were fully empowered to reach definitive and binding agreements which would commit their countries to new investment plans. So *had* been the terrestrial carriers, up to this juncture. Although the 1934 Communications Act did not bestow upon the FCC "any right or responsibility for the negotiation of international facilities engaged in between U.S. carriers and foreign telecommunications entities" (Ibid.: 139), the Commission was now suddenly prepared to intervene in this area. With the advent of satellites, Wiley explained (Ibid.), "it became apparent for a number of reasons that more extensive involvement of the regulatory agency would be needed in the facilities planning and authorization process to assure that carrier decisions were in the public's best interest."

The highly unusual result was that an activist regulatory commission took pains to involve itself in an area long dominated by the very groups it purported to monitor and guide, tying its acquiescence to new capital investment to conditions. Specifically, the Commission stipulated that it would only agree to authorize TAT 5 if appropriate guarantees demonstrated that the cable would be operational during the first quarter of 1970; if all entities sharing in ownership rights in the cable agreed to lower rates for telephone message service and for leased service by 25 percent from previous levels; and if all such entities would likewise agree that satellite and cable circuits should be filled or used at the same proportionate rate—that is, that the unfilled capacity of satellites would be leased at a rate commensurate to that at which the TAT 5 cable would be filled, so that both types of facilities reached the one hundred percent fill figure at about the same time (13 FCC 2d 235 (1968): 1).

This was the Commission's first attempt to assure that newer and potentially cheaper satellite circuits would be utilized, even by carriers and PTTs "whose principal economic and operational interests lay with the older cable technology" (Statement Of Richard E. Wiley *International Telecommunications Policies:* 116). It represents a major commitment to satellites and their prospective users, in that it went far to engage the FCC in crucial and delicate international relations even over and against the interests of the terrestrial carriers. In erstwhile Chairman Wiley's view, "(i)t signalled the FCC's intention to play an active role in the facilities decisions" (Ibid.).

The terrestrial carriers themselves have bitterly protested FCC involvement in facilities planning, characterizing it as "overregulation, which has been the source of unpleasant and potentially serious confrontations between our Government agencies and foreign telecommunications administrations" (Ibid.: Statement Of George F. Knapp, 18–19; Statement Of Eugene F. Murphy, RCA, Ibid.: 2–4; Statement Of Edward A. Gallagher, WUI, 37–51). A recent ally of the carriers is Senator Barry Goldwater, who in 1981 excoriated the FCC for becoming entwined in "a permanent process involving the FCC in discussions and decisions properly left to those private corporations responsible for providing the capital to construct the facilities" (Opening Statement By Senator Goldwater, *International Record Carrier Competition Act Of 1981* Hearing Before The Subcommittee On Communications Of The Committee On Commerce, Science, And Transportation United States Senate 97th Congress 1st Session On S. 271, Serial No. 97-5, 18 February 1981, USGPO: 2).

Complaints by terrestrial carriers over FCC involvement in matters best left to the private sector, however, pale when placed next to the reaction of overseas PTTs. Not surprisingly, the latter "were to consider the proportionate fill requirements as a heavy-handed effort by the FCC to impose its policy decisions upon sovereign nations, and unnecessarily to delay decisions that, from the PTT's standpoint, had already been made in conjunction with the U.S. carriers" (Statement Of Richard E. Wiley *International Telecommunications Policies:* 116). FCC intervention into facilities planning, in Wiley's view, "was to be the start of major problems with foreign PTTs" (Ibid.). Moreover, disagreement between PTTs on one side, and the FCC on the other, has erupted continually since the 1968 proceeding. When the Commission denied the TAT 7 cable plans proposed by the carriers and their PTT correspondents, for instance, James Hodgson—head of the British Post Office and British representative to CEPT—angrily notified FCC Chairman Charles Ferris that, in conjunction with the American "monopoly" of the space segment of Intelsat, the denial of the TAT 7 project in effect comprised an FCC decree in favor of "a U.S. monopoly of the basic means of all intercontinental communications" which "will not go unnoticed in other countries'' (Letter of James Hodgson to Charles D. Ferris, 25 November 1977, FCC Docket No. 18875 "In the Matter Of Policy To Be Followed In Future Licensing Of Facilities For Overseas Communications"). Hodgson said he would study the FCC's record of deliberation upon TAT 7 with care, "both within the British Post Office and in concert with international colleagues." As he characterized it,

> One of the questions which will inevitably arise is how one can reasonably conduct the business of international telecommunications in the North Atlantic area when alone among the community of nations involved one regulatory agency sets itself up as the sole and ultimate arbiter of service, operational, commercial and financial considerations.

Despite the strident opposition of European PTTs to Commission attempts to inject itself into facilities planning—on the side of U.S. dominated satellite circuits (indeed, so much so that the U.S. Department of Defense has felt it necessary to castigate the agency for not heeding the military's requirements for "diversity and redundancy in international communications services" or, in other words, authorization of *both* satellites *and* cables (FCC Docket No. 18875 Ibid. "Petition For Reconsideration Of The Department Of Defense," 2 February 1978))—the FCC has braved hostilities over some fifteen years' worth of deliberations involving four different transatlantic cables. Why?

Robert R. Bruce, formerly General Counsel of the Federal Communications Commission, has tellingly answered this question in recent congressional hearings. The FCC, he stated, "should take the overall policies of our telecommunications partners into account when it considers the authorization of major facilities or individual circuits" (Statement Of Robert R. Bruce, *International Communications Reorganization Act Of 1981* Hearings Before A Subcommittee Of The Committee On Government Operations, House Of Representatives 97th Congress 1st Session On H.R. 1957, 31 March and 2 April 1981, USGPO: 55). It was for this reason, Bruce continued, "that in the past the Commission strongly opposed some legislative and executive branch proposals to remove entirely existing requirements for prior approval of overseas facilities investments.

> The ability to say 'no' to a construction project favored by overseas telecommunications entities is one of the few real points of leverage available to the U.S. in our international telecommunications relations. That leverage should not be abandoned . . . (Ibid.: 56).

In other words, not just satellite expansion but general extension of U.S. domestic and international telecommunications goals should be linked to the facilities planning process—for here, at least, is a real point of leverage. Bruce was not unique in his recognition of the special benefits available to U.S. policymakers through the facilities planning process. FCC Commissioner Abbott Washburn had previously made an analogous point within a more specific context, in his separate comment on the Commission's decision to authorize TAT 7: "Coupled with the need for circuit planning is the necessity for the foreign correspondents to recognize our competitive policies.... Now that the Commission has considered and, to some degree, deferred to the policies and expressed needs of the foreign governments, we should expect 'tit for TAT' from the Europeans in terms of *their* recognition of *our* competitive policies ... Without such reciprocity, I cannot see how we can give any weight or consideration to the Europeans' internal needs and policies in our future dealings" (71 FCC 2d 99, Separate Statement Of Commissioner Abbott Washburn, original emphasis.).

Comsat's dominant role in the Intelsat system thus furnished an opening wedge with which to gain added benefits in the form of concessions from other nations to match newly competitive U.S. domestic policies. Yet this does not by any means exhaust the uses to which satellites may be put. Authorization of domestic satellites within the United States, referred to in Part One, may be singled out as of decisive importance in gaining further international advantages for the transnational corporate sector.

The technical capacities of satellites are still being rapidly developed. "Digital satellite communication networks with ground facilities on customers' premises are expected to become the major medium for long-distance communications for very large organizations," notes one writer (Kaplan, 1979: 42). Private-line traffic growth rates for U.S. domestic satellites over the next decade are estimated at 50 percent a year (Bargellini, 1979: 35). Authorization of many new domestic satellites by FCC regulatory authorities—as discussed in Part One above—is intended to satisfy this corporate user need.

Inevitably, however, it also helps to create great pressure to allow transnational users of U.S. domestic satellite service to extend their use of satellites to varied foreign locations. The example of Satellite Business Systems (SBS), the consortium formed by IBM, Comsat and Aetna Life and Casualty, is illustrative. "Most potential customers for SBS operate in several different countries, if not different continents," writes *Intermedia* ("USA: the ambitions of SBS," 1981: 3): "They will not be satisfied to integrate nationally. They want to integrate internationally." SBS has thus already applied for authority from the FCC to provide private communications services between designated points in Canada and the United States. Such an extension would mark the first transnational operation of private digital satellites, and would require dual approval by U.S. and Canadian authorities. SBS told the Commission in its application for authorization that prospective enlargement of satellite service was contemplated on behalf of five customers hoping to interconnect U.S. and Canadian branch operations: General Motors, ISA Communications Services, Travellers Insurance, Wells Fargo and Westinghouse ("SBS Seeks Authority to Provide Service between U.S. and Canada," 1981: 22). In response the FCC has conditionally authorized both SBS and American Satellite Company to extend the domestic private integrated digital networks of their customers to various Canadian points—"upon coordination by the applicants with Intelsat, and their receipt of intergovernmental concurrence" ("Transborder Telecommunications," 1981: 13). Yet more striking, SBS and British Telecom International have announced plans for the joint provision of transatlantic telecommunications services, for interconnection via satellite of private U.S. networks with British Telecom facilities beginning as early as the end of 1982 ("SBS, British Telecom Int'l Tell Joint Satcom Plans," 1982: 47). SBS has also held discussions with other European PTTs "at the request of its customers" ("IBM and British Firm Discuss Venture To Offer Satellite for Business in Europe," 1981: 30).

Rather than transferring data and other traffic by turns from one nation's terrestrial network to another's, satellites may increasingly allow large users to utilize a global system facility. But does the Intelsat system not already provide such service?

Intelsat apparently suffers from key deficiencies, in the eyes of large corporate users. It is not as advanced technically as the SBS system, although system development is certain to proceed so as eventually to permit Intelsat similar capabilities. However, Intelsat earth stations are as a matter of national policy centralized under national control by PTT authorities in most countries. Given their preference for cables over satellites, PTTs have long tried to integrate Intelsat with their ensconced domestic terrestrial networks (as for that matter has also occurred within the U.S.). SBS and other U.S. domestic satellite systems, on the other hand, aim to confer operating control on large private users themselves: to expedite end-to-end satellite service to corporate customers, SBS typically locates expensive earth station facilities directly upon customer premises. The difference, in sum, is that while the Intelsat facility has been integrated into a hundred-odd national communication systems, SBS's will be integrated directly into private networks. The implications of this distinction are of the profoundest consequence, and have not been lost on telecommunications authorities around the globe. In an era of heightened concern by many countries over extensive transborder data flows eluding national oversight and control, customer-premised satellite earth stations offer corporate users technical means by which entirely to circumvent any given system of national constraints. "What impact do you think new satellite technology will have on the ability of nations and private enterprise to bypass constraints imposed by individual countries (on international data flows—DS)?" former congressman Richard Preyer asked William E. Colby (William E. Colby Responses to Congressman Preyer's Questions, Hearings Before A Subcommittee Of The Committee On Government Operations, *International Data Flow*, U.S. House Of Representatives, 96th Congress 2d Session, 10 13, 27 March; And 21 April 1980: 210). Colby replied (Ibid.: 210-211):

> From a technological point of view, the satellites will have the ability to give nations and even private enterprises an ability to get around national con-

straints . . . . we are apt to see substantial conflict arise between the desires and interests of the individual countries and the sponsors of the industry and the industry itself.

With continuing advance in satellite technics, and with increasing reliance upon such customer-premised earth stations, the balance of power between transnational corporate users and PTT authorities responsible for telecommunications within the nations where transnational users have interests, can be expected to tilt decisively toward the former.

Prospects for private, user-managed satellite networks, originating within the U.S. domestic market, have begun to excite overseas interests as well. It is impossible to predict the outcome of the scramble that is now taking place among an expanding group of satellite equipment manufacturers, financiers, and potential users of satellite services for a foothold in global satellite markets. What *is* predictable is that the general result of the rush to benefit from satellites will leave an ineradicable mark on the nature of telecommunications institutions worldwide. British Aerospace, IBM and British Telecom are pondering plans to start an advanced satellite service linking business customers through Western Europe—which would further accommodate U.S. transnational interests with the British state (De Jonquieres, 1981: 1).* Such a partnership, notes a British business journal, might be used by IBM "as a Trojan horse to conquer the about-to-boom EEC market in telematics" ("Britain bids to breach Europe's telecommonopolies," 1981: 71–72). Comsat, meanwhile, IBM's partner in the SBS venture, has announced that its international unit Comsat General will form a joint venture with British Aerospace in anticipation of the award of a contract to provide leased satellite services to the U.K.'s Ministry of Defence. It is likely, according to Comsat General, "that future defense communications envisioned by the joint venture would result in a substantial increase in the UK aerospace industry's participation in the design and construction of major satellite programs throughout the world" ("Joint Venture Formed for UK Leased Service," 1981: 7). Comsat General is already under contract to the Royal Navy to supply a defense communications system through existing Marisat satellites (Donne, 1981: Financial Times Survey Communications II). Another consortium reflecting British private interests has also already been active in attempting to enter the satellite services market there—this one between British Petroleum, Barclays Merchant Bank and Cable and Wireless. The intensity of activity in the business satellite field in the U.K. in part reflects the preponderance of North Atlantic satellite traffic within the Intelsat system. While the two busiest Intelsat earth stations are located within the United States and account for about 20.8 percent of Intelsat satellite circuits, the third and fourth busiest are in the United Kingdom and together comprise not less than 12.8 percent of total Intelsat circuits ("Twenty Busiest

---

*I should like to thank Janet Wasko for several of the references in this paragraph.

Earth Stations (as of Dec. 31, 1980)," "Comsat Guide to the Intelsat, Marisat and Comstar Satellite Systems," Office of Public Affairs, Comsat: 27). But increasing private-sector participation in the satellite field, whether result of or response to U.S. satellite development, is not limited to the United Kingdom.

Arianespace, a commercial organization comprised of 36 European aerospace companies, 11 European banks and the French Space Agency, is scheduled to obtain private operating control over rocket launching facilities in 1983. The success of the Ariane rocket launched in June of 1981—a project funded largely by France (64 percent) and West Germany (20 percent)—will open the way "to European commercial exploitation of space" ("2 Satellites Placed in Orbit by 10 W. European Nations," 1981: 3A). Internationally the trend follows contours first set down within the U.S. market, as various equipment and service suppliers join with prospective business users in eager anticipation of further satellite development. A thoroughly unprecedented wave of privatization of international telecommunications will likely follow—although whether the move will be spearheaded by U.S. or foreign interests cannot yet be determined. It is, in either case, a reaction to forces first set in motion within the United States.

Within the U.S., these forces continue to be encouraged by Government regulatory authorities. Comsat—the official U.S. signatory to Intelsat *and* the partner of IBM in the SBS venture—has been given the go-ahead for separate, private development of new satellite services and systems by the FCC. Through its role in Inmarsat—the International Maritime Satellite organization—satellite services may be furnished by Comsat *directly* to final users' ships at sea and offshore oil drilling facilities (FCC Docket No. 79-266, "In The Matter Of Comsat Study—Implementation Of Section 505 Of The International Maritime Satellite Telecommunications Act," Final Report And Order Released 1 May 1980 at Paragraph 511). This is in sharp contrast with Comsat's Intelsat role where, recall, it is currently required to furnish satellite service almost exclusively to the terrestrial carriers so that they may resell it to users. Yet the Commission is also investigating benefits contingent on rescinding this "authorized user" policy. Such benefits, according to then FCC Chairman Charles Ferris, would mean that "(u)sers would be able to pay a price based entirely on the advantages of satellite communications. This would also mean that other carriers will compete knowing that if their prices for the same services are set too high, users can go directly to Comsat" (Separate Statement Of Charles D. Ferris Re: International Telecommunications Competition, Ibid.). In a separate proceeding, the FCC heard from William L. Dunn, writing for the *Wall Street Journal* and its Dow Jones parent (FCC Docket No. 80-170, "In Re Petition Of Aeronautical Radio, Inc. For Declaratory Ruling Or Rulemaking That It Is An Authorized User Of The International Telecommunications Facilities Provided By The Communications Satellite Corporation Under The Communications Satellite Act Of 1962," Letter Of William L. Dunn, 6 August 1980):

> As the provider of international and national business and financial news worldwide, it is readily apparent to us that existing constraints in the Authorized User Policies promulgated by the FCC severely impede the financial ability of news disseminators to apply satellite technologies in international operations. The high cost of international communications services as well as the failure of the International Record Carriers to offer unique and innovative services designed to meet needs such as ours have effectively blocked the international application and expansion of news gathering and news distribution techniques commonly utilized domestically. Dow Jones is convinced that international communications costs will be significantly reduced by permitting Comsat to provide service directly to users and by abandoning past ratemaking practices of averaging costs of international satellite and cable services.

Such rate averaging, of course, was designed to placate both the terrestrial carriers and their foreign PTT correspondents who preferred cable to satellite circuits. In the current satellite free-for-all, however, new pressures have emerged within the United States policymaking community to serve major users' needs by eliminating the rate-averaging provision. Another glimpse of the nature of these pressures comes from the Securities Industry Automation Corporation (Ibid., Reply Of 7 October 1980: 3):

> SIAC submits that sound public policy mandates the removal of artificial barriers to competition such as those which currently prevent Comsat from providing international satellite service to the public at large.
>   SIAC's experience . . . demonstrates that both the users and the carriers benefit from increased flexibility in the terms and conditions under which telecommunications services are available. International communications have lagged considerably behind the domestic market because of the difficulty in establishing true competition in that area. Some degree of competition would be encouraged, however, if satellite and cable technologies were separated for private line tariff purposes. In this fashion, the two technologies would compete with one another . . . .
>   SIAC is aware, of course, that the policies adopted in the United States have direct application to only half of the circuit. The control of the other half by foreign administrations can still have a chilling effect on the benefits promised by the Commission's pro-competitive and pro-user policies. Accordingly, SIAC urges the Commission to use every effort to induce the foreign administrations to adopt more enlightened policies.

Thus the "open skies" policy implemented by U.S. regulators in the early 1970s is being rapidly readied for international application—with direct consequences both for Intelsat and for PTT controls over national telecommunications, End-to-end satellite service will allow greater user influence over telematics networks—at the expense of integrated national guidance by PTT ministries. In the international resale and sharing imbroglio, we may glimpse the tensions and antagonisms that are inherent in this expansionary bias of liberalized U.S. policies.

Vans and Resale and Sharing

Mounting strain between U.S. policymakers and overseas PTT authorities is nowhere more palpable than in the results of the FCC's move to authorize international value-added networks. Value added networks (VANs), recall, had already been permitted to operate within the U.S., where they offered facsmile (Graphnet) and packet-switched (Telenet, Tymnet) services. In 1976 Graphnet and Telenet filed applications—except for WUI, the first applications since 1934—for authority to furnish new *international* data and facsimile services (Graphnet Systems et. al., FCC 77-2, released 11 January 1977, 63 FCC 2d 402). Graphnet proposed to extend its domestic facsimile network to international points; Telenet wished to extend its domestic, packet-switched data network to the United Kingdom. Again there is the trend pinpointed by Robert R. Bruce—then General Counsel for the FCC—whereby "competition domestically will result in many entities seeking to provide new and innovative services overseas as well" (Statement Of Robert R. Bruce, Hearings Before The Subcommittee On Communications Of The Committee On Interstate And Foreign Commerce, U.S. House Of Representatives, 96th Congress 1st Session On H.R. 3333 Titles I and III General Provisions, Telecommunication Carrier Regulation, *The Communications Act Of 1979* Vol I Part 1, 24,25,26 April; 1,2,3,4, 8 May 1979, Serial No. 96-121, USGPO 1980: 302).

The logic behind the change in policy followed a now-familiar curve, as then Chairman Wiley explained (Statement Of Richard E. Wiley, *International Telecommunications Policies*, Hearings Before The Subcommittee On Communications Of The Committee On Commerce, Science, And Transportation U.S. Senate 95th Congress 1st Session, 13 July 1977, Serial No. 95-54, USGPO 1978: 108):

> In reviewing these applications, the Commission noted that existing IRC services did not adequately meet the needs of users with intermittent data communications requirements. In many instances the existing IRC data-type services were not cost-effective because of their emphasis on high-volume transmission needs, rather than the short burst and periodic transmission requirements typical of the new data communications activities occurring domestically.

VANs promised to bridge the gap between current record services and advanced data communications. The latter was fast becoming of underlying importance to transnational corporate activities around the globe (Ibid.):

This limitation on data communications developments was viewed as particularly serious in light of the increasing need for domestic users to access computers located overseas to retrieve or input information. Availability of efficient communications capabilities for such database services was, and is, seen as essential since there are located in the United States and the various nations of the world computers which contain large storehouses of information such as technical or medical data of use to businesses or institutions located in another nation. Linking world-wide data bases could lead the way to private and public benefits of enormous commercial and social consequences.

Because the VAN applications would "provide users with the type of data communications capabilities which were not being provided in a cost-effective manner by existing IRC services," the FCC found means of circumventing existing strictures like the international/domestic dichotomy mandated by Section 222 of the Communications Act, and authorized the new VANs to provide service.

At this juncture the distinct stamp of international telecommunications relationships became apparent: the PTTs refused to grant operating agreements to Graphnet and Telenet, which were thus barred from furnishing direct international services. As a matter of policy the European association of PTTs, CEPT, opposes any of its member PTTs recognizing and dealing directly with additional U.S. carriers beside AT&T and the IRCs (Statement Of Philip M. Walker, Hearings Before The Subcommittee On Communications Of The Committee On Commerce, Science, And Transportation U.S. Senate 97th Congress 1st Session On S. 271 To Repeal Section 222 Of The Communications Act Of 1934, *International Record Carrier Competition Act Of 1981* 18 February 1981 Serial No. 97-5, USGPO 1981: 36). As of Spring, 1981, therefore—some four years after FCC authority had been granted—the new VANS were still being forced to offer their innovative services indirectly, by means of interconnection with the IRCs. It may be noted that this delay suits both the PTTs and the established carriers. In Telenet's eyes, anyway, it seems that "most of the foreign administrations have long standing relationships with AT&T and the established international record carriers"—which both are unhappy to relinquish (Ibid.).

A revealing glimpse of the bases for PTT reluctance to grant the requisite operating agreements to the new carriers is found in a discussion offered by GTE Telenet itself (FCC Docket No. 19660, "In the Matter Of International Record Carriers' Scope Of Operations," Rulemaking 690 8 January 1979 "Comments And Reply Comments Of Telenet Communications Corporation": 5). To begin with, the PTTs are government-owned monopolies "whose view as to the efficacy of competition in communications services is substantially different from that which prevails in the United States." Specifically, "considerations other than provision of efficient and economical communications affect the decisions of the foreign administrations." Prominent among such factors has been a "wish to encourage the development of a strong local computer services industry, and, to the extent that terminals in such countries can efficiently and economically access the extensive computer and data base resources in the United States, the development of these resources in such countries might be discouraged." Conversely, Telenet appended, "it is in the U.S.' national interest to encourage the overseas marketing of our computer and data base services." Thus it is a frankly economic conflict which divides PTTs and U.S. policymakers. As Telenet views it (Ibid.: 6):

> In any event, and regardless of the reasons for the unwillingness of foreign administrations to enter into agreements with U.S. entities seeking to provide new or additional communications services, the effect of that refusal is that the Commission's ability to influence the availability of international services is greatly impaired.

The Commission's ability to influence the course of international telecommunications policy was crucial, however, when the larger interests of transnational corporate users were taken into account. Robert R. Bruce of the FCC recognized this when he argued that the refusal of the PTTs to grant operating agreements to the value-added carriers "has limited more than U.S. consumers' choices among overseas carriers . . . . In this area, international communications policy can have a broad impact on U.S. overseas business activities and on the balance of payments." Therefore, he announced, "(t)his problem simply cannot be ignored"; "(w)e obviously have to be able to work with foreign interests but we need some leverage in that process" (Statement Of Robert R. Bruce, *The Communications Act Of 1979*: 303–304, 322).

One form of leverage utilized by the FCC in such circumstances has already been outlined in discussion of the facilities planning process. By this means, FCC authorization of new international cable circuits was linked to PTT willingness to liberalize constraints on international services. A second kind of leverage, closely related to facilities planning, may be obtained through what Robert R. Bruce calls "upgrading priorities, increased policy focus, and better and higher level coordination" (Statement Of Robert R. Bruce, Hearings Before A Subcommittee Of The Committee On Government Operations, *International Communications Reorganization Act Of 1981*, U.S. House Of Representatives 97th Congress 1st Session On H.R. 1957, 31 March and 2 April 1981, USGPO 1981: 54). In other words, relations between the FCC and other federal agencies and private sector representatives should be characterized by what Commissioner Robert E. Lee labeled "constant contact" (Ibid.: Statement Of Robert E. Lee, 195). The consultative process by which the Commission now attempts to coordinate international facilities planning involves extensive information gathering, in turn facilitated by reliance upon several ad hoc groups chaired by representatives of terrestrial carriers, Comsat, and federal agencies. Members of the FCC participate in these groups, yet the overall planning effort embraces the State Department and the National Telecommunications and Information Administration as well. These complex procedures, Lee emphasized (Ibid.), "lead to a continuing exchange of information and coordination of views among the federal agencies and the private sector" and now, still according to Lee, better take into account "the needs of the user community" (Ibid.: 195 and 191). Moreover, the role of the "user community" in international policy deliberations is being enlarged and cultivated by Government officials themselves. In 1980 hearings, for example, a subcommittee of the Government Operations Committee of the House solicitously inquired of the Control Data Corporation, "What can the government do to better ensure that the interests of the users of international telecommunications services are represented when negotiations with foreign PTTs take place?" ("Responses To Questions Submitted To The Control Data Corporation," Hearings Before A Subcommittee Of The Committee On Government Operations, U.S. House Of Representatives 96th Congress 2d Session, 10,13,27 March; 21 April 1980, *International Data Flow* USGPO 1980: 55).

The consequence of enlarged sensitivity by U.S. authorities to the needs and interests of transnational corporate users of telematics has been an escalation of tension between U.S. policymakers and PTT ministries over continuing attempts to extend "procompetitive' policies into the international sphere. An outgrowth of the VAN authorization decision, having to do with international resale and sharing of private line circuits, makes this quite clear.

To comprehend the conflict over resale and sharing we must step back for a moment to recall the broader framework in which deliberations over international telecommunications policy are now taking place. Users need and want a globe-encircling capacity for data communications, because only this will permit worldwide coordination of transnational business. Data communications cannot develop, however, without massive upgrading of world telecommunications facilities. Computer communications—as we have seen at length in Part One— harbors quite different technical requirements than those of conventional voice telephony. Connection times, line speeds and error rates needed for efficient computer-to-computer "talk" are far out of reach of the unconditioned telephone network. For this reason, major users have increasingly relied on dedicated private line circuits. Leased at flat monthly rates, private lines are desirable both on grounds of confidentiality and cost, and also because they may be specially customized and conditioned to meet particular user needs.

In many nations apart from the United States, however, the government ministry of posts, telephones and telegraphs which owns and operates communications network facilities within its borders, quite naturally seeks to direct further technical advance by developing data communications networks as public enterprises under its own guidance. Some 16 European PTTs, in fact, are now rapidly developing public packet-switched networks.

To encourage use of these public networks PTTs apparently are pondering proposals which threaten to eliminate private line service (Statement Of John Eger, *Hearings On International Data Flow* 1980: 177). U.S. transnational users thus worry that PTTs may price private line facilities "at rates and schedules that are prohibitive for the development of private or user-controlled networks" (Ibid.). Added incentive to eliminate private lines comes from the convergence of telecommunications and the computer referred to above: when private lines are interconnected across national boundaries, computers in one country may effectively perform work that might have been done in another. Such a turn of events, of course, would affect and probably damage larger foreign national policies aimed at bolstering domestic economic growth through encouragement of a native computer and data processing industry. From the U.S. users' perspective, on the other hand, escalating communications costs, redundancy of computer facilities and possible service problems attendant on shifting from customized private lines to "all purpose" public networks are alike critical drawbacks of such plans.

The capacity for resale and sharing impinges directly on this larger concern. When users of private lines (or of WATS or MTS) are allowed to share these circuits with others, or to resell unused circuit capacity—both of which have become economically attractive in development of "demand-oriented networks"—users enter the communications business. Supporters of this move assert that it depresses prices to reflect costs, as common carriers must compete with private resellers, that it helps to eliminate cross subsidies between services and adds incentive to introduce new services tailored to dynamic user needs.

Neither carriers nor overseas correspondent PTTs, however, held a desire to bring resale and sharing into international services, for the very same reasons that users find them attractive: private lines are leased at flat monthly rates and by increasing their use through resale or sharing potential public-network traffic may be diverted to private resellers. This may in turn jeopardize the tariff structure of a given PTT. The only reason international private lines have even been tolerated thus far, according to *Business Week* ("An FCC Plan Backfires Abroad," 1980: 56), is that "regulations exist at both ends of the circuit to prevent resale and sharing."

The FCC, hoping to extend newly liberal domestic tariff offerings overseas, struck directly at these restraints. In the original *domestic* resale and sharing proceeding, in fact, over and above arguments by RCA and ITT that underscored the PTT's pointed opposition to resale and sharing, the Commission reached the following striking conclusion (FCC Docket No. 20097, "In The Matter Of Regulatory Policies Concerning Resale And Shared Use Of Common Carrier Services And Facilities," Report And Order 16 July 1976, 60 FCC 2d 298):

> We have considered the views that the IRCs should not be subject to our policy regarding resale and sharing of their services, or, in the alternative, that resale and sharing should be required only if the foreign entity agrees to such practices. Neither position is, in our view, consistent with the statutory obligation of the IRCs as common carriers subject to our jurisdiction . . . . We may not exempt the IRCs from bringing their tariffs into agreement with our

interpretation of the law simply because a foreign entity does not agree with us, for whatever reason . . . our jurisdiction over the charges (and practices) of the IRCs is not dependent upon the agreement of their foreign correspondents.

Six months later, in response to petitions for reconsideration filed by WUI and TRT, the FCC reversed itself (Memorandum Opinion And Order, 12 January 1977, 62 FCC 2d 593). Choosing not to discuss substantive issues at that time, the Commission determined simply that "our policy requiring unlimited resale and sharing of private line services and facilities should not at this time be applied to international services." A separate proceeding to be instituted would address the issues, and the FCC stressed that "our action herein is not a finding that existing restrictions on international services and facilities are just, reasonable or not unlawfully discriminatory" (Ibid.).

In the constant contest that international telematics policymaking has become, it cannot be said whether this momentary aggression comprised thrust or parry. What may be asserted without fear of question is that this minor skirmish was soon succeeded by a more aggravated engagement. In the Spring of 1980, the Commission issued a proposal to extend resale and sharing into the international arena (FCC Docket No. 80-176, Notice Of Proposed Rulemaking 19 May 1980). Virtually at once, a strongly worded letter was sent to the State Department, which relayed it to the FCC, by the Director of the International Telecommunication Union's International Telegraph and Telephone Consultative Committee (CCITT). The ITU is the 155-nation agency that coordinates worldwide telecommunications; the CCITT issues nonbinding but routinely accepted "recommendations" on use of international telecommunications. The FCC proposal, wrote Leon Burtz (Ibid., 20 June 1980), had engendered "surprise" and "deep disappointment" within the ITU. "It seems to me an extremely dangerous situation," Burtz charged, "when one country, and what is more, the leading country with regard to the number of subscribers, the extent of its services and its telecommunications technology, can help to undermine the work of the CCITT."

It was not the CCITT alone that reacted to the FCC. Telegrams were received from two dozen countries responding to the decision (Appendix To Comments Of WUI, 15 August 1980, Ibid.). Their messages are succinct: "On behalf of the Belgian PTT, this is to inform you that there is no intention to deviate from the binding rule"; "we take this opportunity to officially restate . . . our firm opposition to the concepts of Shared Use and Resale" (Italy); "Any major changes to the conditions of use applied to international leased circuits . . . . must necessarily be preceded by full discussions in the ITU" (UK). A copy of a telegram from Kuwait to the ITU Secretary General, received by the State Department, openly voiced the threat underlying these comments (Geneva 12935; 14 September 1980, 051813Z). The ITU should "initiate necessary action for reversing the FCC decision," asserted Kuwait, and if appeals and persuasion fail, drastic action is called for from the international telecommunications community by abolishing international private leased circuits. By this move, those multinationals for whose benefit FCC is making the decision, will end up by paying 5 to 10 times more as they will be compelled to route their traffic via international public data networks . . .

As was later revealed, ongoing discussion within the CCITT already included matters "of direct relevance to the FCC proposal"—in the form of an Italian suggestion that private line flat-rate tariffs should be replaced by usage-sensitive pricing (FCC Docket No. 80-176, Comments Of The International Communications Association, 7 August 1980: 10). Should this occur the heaviest users of communications circuits would experience large cost increases. The FCC apparently had taken unilateral action on this delicate issue, ignorant of or indifferent to the fact that discussions of the same sensitive matter were simultaneously occurring in a high-level international forum. The Commission therefore both confused and angered foreign authorities while serving notice on the international telecommunications community that U.S. policymaking in this field was uncoordinated and fragmented.

Among U.S. transnational corporate users, reaction to the FCC was swift. If PTTs phased out private lines, stated one journal ("An FCC Plan Backfires Abroad," 1980: 56), "the impact on some U.S. companies could be devastating." Philip Onstad of Control Data Corporation estimated that for his company such a move would result in directly increased costs of 700 percent (Ibid.). Because public data networks had neither the flexibility nor the customized character of leased lines, moreover, Onstad worried that "our entire communications systems would be downgraded."

Another major user of private lines, Citicorp, wrote the FCC to assert that (FCC Docket No. 80-176, Comments Of Citicorp 15 August 1980 Ibid.):

> The loss or restructuring of these essential facilities would have a dramatic adverse effect upon the American banking community and would raise the spectre of limitations on commercial access to foreign markets. . .The real and theoretical benefits which may flow from implementation of the Commission's proposal would be a small victory indeed if the unilateral adoption of sharing and resale precipitates foreign limitations on or rate restructuring of existing private line facilities.

PTT control of communications circuits at the foreign end thus forced transnational users of private lines to demand, albeit possibly unhappily, that the FCC reconsider its proposal. The International Communications Association informed the FCC that, "(i)nasmuch as a primary objective of United States international telecommunications policy must be continued availability of leased lines, the development of international regulatory policy must reflect a realistic appraisal of positions taken by foreign administrations in respect of such services and a keen awareness of user requirements" (Ibid.: Comments of ICA 2). The Commission thus "should not even suggest support for usage-sensitive pricing without an in-depth study as to its practical effects" (Ibid.: 14). More broadly, the resale and sharing debacle had demonstrated that, clearly, the coordination and control of U.S. international telecommunications policymaking was insufficiently responsive to the largest users' interests. "(I)nternational telecommunications policy cannot be made," declared the Chase Manhattan Bank (Ibid.: 14 August 1980), "without assessing the whole of our relationships with other nations; nor it would seem to follow, making any significant changes in our policy which would impact those of others, without full discussion and adequate consultation and achievement of consensus on a bilateral or multilateral basis, as the issue requires.

> Toward that end, a necessary first step might entail a total review and evaluation with the Department of State, National Telecommunications and Information Administration (NTIA), and other affected or interested government agencies of the multifaceted interests which the U.S. has at stake, taking into account, of course, the broad interests of major international users, such as CHASE.

More than any other recent proceeding the resale and sharing debacle displays the zealousness with which the FCC has accepted its charge as a lead agency, acting largely—though by no means only—on behalf of transnational corporate users in the international information war. Yet even as these large users tried to moderate the FCC's approach, the NTIA continued to hammer out the hard line (Ibid.: 15 August 1980: 15-16):

> While the United States should give due weight to the CCITT recommendations, we do not believe so fundamental a national goal as the encouragement of competition and market entry should be frustrated by international recommendations.

"The rapid explosion of telecommunications and information technology in the United States, coupled with equally explosive demand therefor," NTIA brusquely continued (Ibid.), "has simply required the development of policies which are not necessarily consistent with those recommended by the CCITT."

And so the pressure continues, and despite the temporary slow-down in the resale and sharing proceeding, continuing advance in provision of services and equipment within the domestic market forces the pace of international telematics development. This occurs not because of some ineluctable economic logic but because the rules guiding telematics have been—and remain—subject to systematic overhaul and transformation by Government authorities. In the United States, as this book has tried to demonstrate, Government has not been oblivious to its task.

Up to this point, however, Government authorities have been comparatively lucky. Conflicts over the integration of telematics into world society give every promise of growing broader, deeper and sharper. Against a backdrop of protectionism, higher unemployment rates, stagflation and a new fragility in postwar political allegiances, unexpected tensions and antagonisms may be expected to emerge from the computerization of society itself.

For let there be no mistake—telematics is today a battlefield on which the underlying shape and imperatives of the entire world economy are being contested. Final results of this international information war are unpredictable, even unforseeable. Yet its implications are starkly clear. Whoever controls the equipment markets, the software and services that are telematics stands to benefit from an unprecedented centralization of control over global economic activities and resources.

This is why the United States Government, like that of other nations, has such a necessary and critical role. The Government has for decades strived to enhance telematics capabilities on behalf of the largest corporate users, at first domestically, then in a global context. More and more responsive to demands voiced by the transnational business sector as a whole, less and less narrowly a mere creature of the regulated industries themselves, however vast their power or resources, Government authorities and, especially, the Federal Communications Commission, have been vital in rapidly infusing telematics equipment and services into the social economy.

Yet even this profoundly important policymaking role does not exhaust the Government's contribution to computerization. In the final Part of this volume I inspect briefly the nature of the Government as a consumer and user of telematics. In short, I turn to scrutinize the Government telematics *market*.

THREE

Telematics and Government

"If the people who make policy control information about making policy, then the rest of us are inconsequential citizens."
———William Appleman Williams, 1981

"The long-term future of the Postal Service should be re-examined. In a competitive age in which the telephone, not the post, provides essential communication, it is unclear why there should be a publicly owned and supported national document delivery company."
———The Heritage Foundation, in Rodgers 1981

THE GOVERNMENT TELEMATICS MARKET

Through research and development funding, setting of various technical standards and contracts for telematics equipment and services, the Federal Government has played a major role in the invention and innovation of telematics in U.S. society. It has been amply noted that Government is "by far the largest purchasor and lessor of computers" (Gilchrist and Wessel, 1972: 11), and that by virtue of prodigious expenditures from World War Two down to the present, Government has proved a formidable source of market power in telematics industries (Brock, 1975: 137). How much does Government spend on computer communications? How extensive is Government influence on the size and scope and character of telematics industries?

One might easily assume that either of these related questions should be simply answered. In fact, it is difficult to supply even approximate answers. Appraisal of the Government telematics market is hindered by several factors. First, cost accounting for telematics equipment and services is hampered by substantial definitional disagreement about what budgetary items are to be encompassed (General Accounting Office, "Accounting For Automatic Data Processing Costs Needs Improvement," Report to the Congress by the Comptroller General of The United States, FGMSD-78-14, 7 February 1978; King and Kraemer, 1981). Second, a very high proportion of Federal telematics outlays derive from military projects cloaked in the usual secrecy. Finally, despite varied moves to centralize procurements in this field and to monitor agency performance, Government utilization of telematics tends to be uncoordinated, even balkanized—perhaps, as we shall see, as a result of deliberate policy. For these reasons even those segments of Government telematics that are not confidential tend to be inaccurately and incompletely measured.

An assessment of at least some key dimensions of the Federal telematics market must nonetheless be attempted. I have assembled the following relevant data:

1) Government funded U.S. electrical equipment and communications industry firms with about $2.7 billion in 1977 for research and development—45 percent of total industry R and D costs (*Research and Development in Industry 1977*, National Science Foundation, USGPO, 1977: 2, 24). In 1980, however, the Research, Development, Testing and Evaluation budget for the Department of Defense alone was $13.5 billion—of which some $6.3 billion was targeted for ra-

dio and television communications equipment (*1981 U.S. Industrial Outlook*, U.S. Department Of Commerce January 1981: 307).

2) Government purchases of electronic equipments and systems in which such equipment is embedded apparently comprise "one-half of the total value of shipments" for this industry; in 1980 this total was estimated at $47.6 billion (Ibid.: 305). Overall U.S. Government investment in communications-electronic equipment has been valued at "approximately 67 billion dollars" (Jansky, 1980: 314). This investment, it may be noted, is about half the size of the Bell System's total assets. Real growth in a submarket of the electronics industry, radio and TV communication equipment, "depends primarily on policy decisions at the national level, since the U.S. Government is the most significant purchaser of its products," notes a Commerce Department source (*1981 U.S. Industrial Outlook* Ibid.: 306). In 1980, Defense Department expenditures in the field by themselves comprised nearly $14.5 billion (Ibid.: 307).

Military use of radio and TV units is bound up closely with a long-term commitment to the "electronic battlefield"—through which computer communications are used to assist coordination and control of eveything from troop movements to missile launches to tank maneuvers. "There is hardly a piece of equipment that I put out that does not have some ADP (automatic data processing—DS) in it," states General Albert N. Stubblebine III, of the Army Electronics Research and Development Command (*Department of Defense Authorization For Appropriations For Fiscal Year 1981*, Hearings Before The Committee On Armed Services, U.S. Senate 96th Congress 2d Session On S.2294, Part 5 Research and Development; 5, 11, 12, 13, 14, 25, 26 March 1980, USGPO 1980: 3040). "Literally every weapons system that we are planning and bringing into development in some way employs minicomputers and microelectronics," adds Lt. General Donald R. Keith, U.S. Army Deputy Chief of Staff for Research, Development and Acquisition (Ibid.: 2762). Dr. Percy A. Pierre—Assistant Secretary of the Army, Research, Development and Acquisition—explained (Ibid.: 3036):

> computers are being distributed throughout the battlefield and we rely on them more or less almost everywhere . . .the philosophy of design has got to be to distribute the computer power throughout the area rather than bring it all into one place and to internet the different computers so that when one goes down another one can help.

Overlap between the top one hundred military contractors and the top one hundred data processing companies in 1979 is shown in Table 12. Twenty firms find their way onto both lists; this marked correspondence *excludes* other prominent Defense contractors in electronics, telecommunications and areospace: Westinghouse, Fairchild Industries, RCA, Ford Aerospace, GTE and Kodak. Too, the

TABLE 12
Top 100 Data Processing Companies and Top 100 DoD Contractors 1979 Overlap

Company	Datamation 100 Ranking	DoD Contractor Ranking
IBM	1	19
Control Data Corporation	4	76
Sperry Rand	5	14
Honeywell	7	17
Hewlett-Packard	8	87
Xerox	12	94
TRW	13	27
Texas Instruments	14	29
Computer Sciences Corp	15	79
General Electric	17	4
ITT	24	37
McDonnell Douglas	25	2
Harris	28	56
Raytheon	36	9
Teletype (AT&T)	41	18
Boeing	57	7
Martin Marietta	67	20
Lear Siegler	70	92
Sun Company	71	73
Grumman	97	8

Sources: "Defense Dept. Lists Top 100," *Aviation Week & Space Technology* 7 July 1980: 64–67; "The Datamation 100," *Datamation* July 1980 (Vol. 26 No. 7): 98–101 (wordwide dp revenue is gauge of dp rank).

Defense Department alone spends more than $2.8 billion annually for telecommunications and command and control programs (Statement of Dr. Richard D. Delauer, Undersecretary of Defense for Research and Engineering, Hearings On S. 898, *The Telecommunications Competition And Deregulation Act Of 1981,* Committee On Commerce, Science And Transportation, U.S. Senate 97th Congress 1st Session, 2,11,15,16,19 June 1981, Serial No. 97–61. Washington: USGPO, 146).

The ever-deepening dependence of the military upon telematics, and the present role of the Department of Defense in helping to shape policy in this field, can be comprehended only if it is recognized that military and civilian communications systems have long been deliberately interwoven. The Federal Government today "obtains more than 94 percent of its most critical domestic communications circuits from commercial carriers" (Ibid.: Statement of LtG. William J. Hilsman, 149). The Strategic Air Command, the Joint Chiefs of Staff, and other top military units rely routinely upon the commercial carriers "as their primary method for command and control" (Ibid.: Statement of Dr. Richard D. Delauer, 146).

Corresponding to this policy of interpenetration of commercial and military facilities, extensive and detailed planning jointly conducted by the telecommunications carriers and the DoD has become a quotidian feature on the telecommunications landscape. Indeed, the importance of this planning effort in the eyes of DoD "cannot be overstated": "we must and do also work closely and cooperatively with telecommunications carriers on a daily basis" (Ibid.: Hilsman, 149, 148).

All of this would be merely routine were it not for the tremendous changes being wrought in the U.S. telecommunications industry. Developments described above have brought in their train new difficulties for military communications network planners and centralized network management. Proliferation of many private networks and emergence of various new telecommunications carriers has motivated DoD to strive for a means of exerting centralized, unified planning and control over the now-fragmented telecommunications system. Rephrased, the gratification of corporate users' telematics demands has incurred anxiety and dissatisfaction within the ranks of the very largest user, DoD.

Owing to the increasing fragmentation and complexity of U.S. telecommunications and the dependence of the military upon private industry for communications, Undersecretary of Defense Delauer explains (Ibid.: Delauer, 146), "in the last decade the Department of Defense has had to become more extensively involved in how the rules and regulations governing the telecommunications industry are developed." This has necessitated "extensive participation in both federal and state regulatory proceedings, addressing not only the rates and charges for telecommunications service, but also the manner and technical conditions under which service is to be provided, and the structure of the industry itself" (Ibid.). Currently, for example, DoD is participating in not less than fifty federal and state regulatory and judicial proceedings in this area (Ibid.: Hilsman, 148). At the broadest level, DoD believes that "the cognizant regulatory agencies remain handicapped by rigid or outmoded laws in addressing the impact of technological and structural changes upon national defense and security" (Ibid.: Delauer, 146). Moreover, military planning has been "rendered extremely difficult" by concurrent efforts to frame policies for the telecommunications industry in the Judiciary (the antitrust suit against AT&T), the regulatory sector (the FCC and State Public Utility Commissions) and the legislature (Ibid: 147). Akin to other corporate users, therefore, but with its own particular interest in central, unified planning, the DoD has actively engaged itself in attempts to reinstate certainty and predictability into U.S. telecommunications. Also like other large users, DoD's attempt is naturally framed so as to be maximally conducive to its own special needs.

Extensive military participation in support of AT&T during the recent antitrust case is one example (Hirsch, "Judge lets AT&T Use DoD Study," 1981: 6). The Defense Department has also entered the legislative fray. To address the extensive and growing interrelationship between the telecommunications industry and military needs, Undersecretary of Defense Delauer asserts, "it is DoD's view that any legislative regulatory reform initiative must explicitly recognize and declare the vital and increasing importance of the Nation's telecommunications resources to meeting such needs" (*Telecommunication Competition and Deregulation Act of 1981*, Delauer: 146). Above and beyond the economic demands made by other corporate telecommunications users, "we at DOD . . .have broader needs to which the telecommunications industry must respond" (Ibid.: Hilsman: 148). The Undersecretary concluded his case for a "framework of certainty" by stating that "Congressional action must override any separate and ongoing, regulatory, or judicial attempts to restructure this critical industry" (Ibid.: 150). The Senate bill—S. 898— should be revised to include greater concern for the special demands of DoD: "(T)here must be pervasive recognition in the statements of findings, purposes, or policy contained in S. 898, that the promotion of national defense and security and emergency preparedness is a goal at least equal to any other" (Ibid.).

As passed by the Senate, S. 898 now includes a provision allowing the president, on a Defense Department recommendation, to require any communications company to furnish any services, facilities or equipment "to promote the national defense and security or the emergency preparedness of the nation;" no state of emergency need be declared for this to occur (Brown, 1981: A1). The acting director of the Systems Evaluation Division of the Federal Emergency Management Agency, A. L. Henrichsen, warned that should this provision be enacted into law, it would grant DoD "statutory authority to impose 'national security' requirements on the telecommunications industry apparently without review except by the courts" (Ibid.). Competitors of the Bell System are disturbed that the measure may be aimed at permitting the military to get around restrictions aimed at limiting AT&T's activities in a deregulated environment. Former FCC Commissioner Kenneth Cox, now vice president of MCI, claims "they have snuck in there some powers that are quite dangerous for us as a company and for the public at large" (Ibid.). Continuing military forays into the field of telecommunications policymaking follow naturally from the fact that "(t)he line between the civilian and the military seems to have grown thin . . ." (Oettinger, 1980: 197); any further escalation of DoD telecommunications policymaking intervention would compel nothing less than a full-fledged military occupation of the field.

3) In 1968, Brock estimates (Brock, 1975: 137), the Government accounted for 17 percent of the total U.S. computer market. As computers became ubiquitous, the Government share declined; a 1979 reckoning found that the Federal share of the U.S. installed computer base was about 6.9%—but exactly what this figure takes into account— lease arrangements? special-purpose systems? commercial services?—is unclear (Standard & Poor's Industry Surveys, *Office Equipment Systems and Services* 15 May 1980: 014).

The Government probably employs between 100,000 and 150,000 technical personnel and computer specialists to operate its vast data processing facilities (Kirchner, "Bad Tidings in Store for Federal Meet Attendees," 1981: 15; General Accounting Office "The Federal Information Processing Standards Program: Many Potential Benefits, Little Progress, And Many Problems," Report to the Congress by the Comptroller General of the United States, FGMSD 78-23, 19 April 1978: 1). The latter embraced 14,333 central processing units as of 1979 (Kirchner, "Federal CPU Inventory UP," 1981: 67–68; "Automatic Data Processing Activities in the United States Government as of the end of Fiscal Year 1979, Summary, General Services Administration, Automated Data and Telecommunications Service September 1980, USGPO Stock No. 7610-00-111-8245: 11). Of these, 9046 are in a "Special Management Classification" category—meaning that they are used in classified, control or mobile environments, and cost and utilization data are not reported. The remainder, numbering 5287, by themselves account for $4.8 billion in annual costs during 1979 (Ibid.: 56); but these, called "General Management" systems, comprise scarcely 37 percent of reported Government systems. If the 63 percent of computer systems for which data are not given are proportionate in cost to the General Management Systems for which cost data are supplied, then operating costs for all Government computers would have been roughly $13 billion in 1979. The General Accounting Office estimated that, in 1977, Government spent "over $10 billion . . .for ADP equipment and technical personnel" (General Accounting Office "The Federal Information Processing Standards Program . . ." 1978: 1).

Another GAO survey, however, found that annual ADP costs might "run as high as $15 billion ("Shifting The Government's Automatic Data Processing Requirements To The Private Sector: Further Study And Better Guidance Needed," Report By The U.S. General Accounting Office, FGMSD-78-22, 11 April 1978: 2).

4) Computer software or programming instructions, may add $6 billion a year to the Federal ADP outlay (Kirchner, "GAO Hits Federal Outlays for Software Upkeep," 1981: 13; "Wider Use Of Better Computer Software Technology Can Improve Management Control And Reduce Costs," Report to the Congress by the Comptroller General of the United States, General Accounting Office, FSMSD-80-30, 29 April 1980: 1). Thus, by 1977, accumulated Federal investment in software was in the neighborhood of $25 billion (Ibid.).

The bulk of Government software expenditures again come from the military, which largely underwrote the now ubiquitous COBOL (Common Business-Oriented Language), in 1959 (Brock, 1975: 144–147). Military expenditures for software may reach $5 billion annually (King and Kraemer, 1981: 127).

Federal outlays cumulatively furnish impressive evidence for a massive Government role as a telematics industry market force. One cannot merely sum up the various figures calculated above; and many other expenditures may simply be absent. Nonetheless, a conservative estimate of total Government telematics spending might include: Defense research and development funds for radio and television communications equipment—$6.3 billion; military equipment expenditures—$14.5 billion; Defense telecommunications and command and control program outlays—$2.8 billion; automatic data processing equipment and technical personnel funds (1977 estimate)—$10 billion; computer software—$6 billion; yielding a grand total of $39.6 billion. If we reduce this figure somewhat to take account of any overlap that might exist in radio and television equipment outlays and ADP equipment spending, it nonetheless seems certain that Government at a minimum furnishes a $30–35 billion telematics market. IBM's total revenues in 1980, by comparison, amounted to $26.2 billion.

Just how dependent are telematics companies, however, on this vast Government market? Although, again, the full extent of corporate dependence may be clouded for a variety of reasons—subcontracting and indirect sales may be far larger than direct sales to agencies and departments—a fairly clear and unambiguous picture emerges. Table 13 shows selected data concerning sales to the Government for the top eight data processing firms (as ranked by *Datamation*) for 1979. The Table should be considered with caution, as it reports data as given in each firm's yearly report to the Securities and Exchange Commission— the 10K Report—and these data may not always be comparable. Digital Equipment Corporation, for instance, reports that "approximately 4% of the company's total sales were made directly to various agencies of the U.S. Government (Digital Equipment Corporation 1980 10K: 3). Hewlett-Packard, however, states that in fiscal 1979 about 12% of its incoming orders "were derived directly or indirectly from the United States Government" (Hewlett-Packard, 1979: 3). IBM, meanwhile, reports only that its Federal Systems Division had 1979 revenues of $612 million—for about 2.7 percent of total revenues—but this estimate excludes sup-

TABLE 13
Top Eight DP Companies and the Government Computer Market 1979*

Company	No. Computers Supplied to Government	Estimated Costs** ($ millions)
IBM	1284	1457
Burroughs	303	309
NCR	—	—
Control Data	497	754
Sperry Rand	1778	687
Digital Equipment	3656	241
Honeywell	896	534
Hewlett-Packard	1093	60

Notes: *Top eight DP firms ranked by worldwide DP revenues 1979 ("The Datamation 100," *Datamation* July 1980 (Vol. 26 No. 7): 98–101.
**Estimated cost of total components, owned and leased.
Source: "Automatic Data Processing Activities in the U.S. Government as of the end of fiscal year 1979, Summary," General Services Administration Automated Data and Telecommunications Service, September 1980, Stock No. 7610-00-111-8245, Washington, D.C. USGPO: 17, 60.

plies and services provided to Government through IBM's Data Processing and Office Products Division (IBM 1979 Annual Report). And while Honeywell found some 16 percent of its total revenues in Government contracts, although not all of these procurements went directly for information systems, Control Data reports that it obtained 11 percent of its computer business revenues alone from the Government—down from 21 percent in 1975 (Honeywell, 1979: 3; Control Data Corporation, 1979: 3). Sperry, finally, records only that during 1979 "revenue under United States defense and space contracts and subcontracts represented approximately 16 percent of the Company's total revenue, and sales of commercial products to the U.S. Government represented an additional 4 percent." Government contracts comprised 21 percent of its computer systems and equipment division's revenues (Sperry Rand, 1979: 3).

Two of the largest computer services firms, Electronic Data Systems (ranked eighteenth among data processing firms by *Datamation* in 1980) and Computer Sciences Corporation (ranked fifteenth), furnish further evidence of the telematics industry-Government tie ("The Datamation 100," 1981: 102–103). Electronic Data Systems, with revenues of $414 million in 1980, contracts to supply and operate systems so that "in effect, (it) becomes that customer's data processing department" (Electronic Data Systems, 1979: 2). Systems and documentation developed by EDS, however, "generally remain its property"—although, when a given contract terminates, EDS may often assist in converting material for customer use "for an additional fee" (Ibid.). (A similar tendency is recorded by the General Accounting Office's untitled publication of 18 May 1979, labeled PSAD-79-69, which details in a letter to the Secretary of Defense from J. H. Stolarow, Director of GAO's Procurement and Systems Acquisition Division, how Government paid $1.858 million to Grumman Data Systems for use of special purpose computer software packages, "even though these packages had been developed primarily at Government expense" (Ibid.: 1). This occurred even though, by Defense Acquisition Regulation 9-602, Government is to have unlimited rights in computer software required to be developed under or generated as a necessary part of performing a Government contract.) About 20 percent of 1979 EDS gross revenues came from Federal, State and Local Governments, the former accounting for 8 percent by itself; an additional 9 percent of revenues derived from services to corporations with prime government contracts (Electronic Data Systems, 1979: 2). In contrast, Computer Sciences Corporation has but one major customer—the Federal Government. Fully 71 percent of revenues came from that source in 1979, with state and local governments contributing an additional 7 percent (Computer Sciences Corporation 1979: 1). Key agencies sustaining CSC have been NASA, which accorded 26 percent of total revenues, the Navy (15 percent) and GSA (13 percent) (Ibid.). CSC's offerings include its worldwide remote INFONET computing service, which makes use of computers in Los Angeles, Chicago, Washington, D.C., Toronto and Calgary (Ibid.: 4). In fiscal year 1980 the firm received 63 percent of total revenues from the Federal Government

("GSA Eases CSC Suspense," 1981: 96); revenues for 1980 came to $560 million.

Yet if the telematics industry seems frequently to rely upon Government contracts for a significant portion of its revenues, a more formidable and far-reaching form of dependence may be that of Government itself upon these same telematics companies. Through a pronounced and deliberate disinclination to impose technical standards for automatic data processing activities, as well as through noncompetitive procurements, Federal agencies "have become locked into suppliers of computers and related services . . . Agencies have become increasingly dependent on single sources of supply for their equipment and software" ("The Federal Information Processing Standards Program: Many Potential Benefits, Little Progress, And Many Problems," Report to the Congress by the Comptroller General of the United States, General Accounting Office FGMSD 78-23, 19 April 1978: 4). "Little progress has been made in the Federal ADP standards program" directed by the National Bureau of Standards, this GAO study charged (Ibid.: 17), "because its managers have depended too much upon the commercial sector to develop standards for Federal use." The National Bureau of Standards—perhaps attempting to get off the hook itself—"has cited the dominance of large manufacturers in the commercial process as a primary reason for the slow progress in developing commercial standards" (Ibid.: 18).

Partly in reaction to this balkanized situation Public Law 89-306—the Brooks Act—was passed in 1965, as, by that year, "the lack of ADP standards was believed to have seriously compromised the Government's overall ADP potential" (Ibid.: 2). The Act therefore called into being a Federal ADP Standards Program that would "permit the interchange of computer equipment, software, and data," and which was equally "intended to stimulate competition by permitting Federal agencies to procure their ADP requirements from numerous vendors offering low-cost compatible products" (Ibid.). Yet, the General Accounting Office reported in its 1978 study (Ibid.: 4), "Contrary to a major objective of the Brooks Act, the Federal Government is not fully realizing potential savings available through competitive procurements." Agencies and Departments have grown dependent on specific suppliers "either because certain essential standards have not been developed or agencies are not complying with existing standards. As a result they are making noncompetitive procurements to avoid extensive efforts to convert their computer programs and data" (Ibid.). Such software conversions—that is, translations of programming instructions from one computer language to another to meet variant hardware demands—by 1977 cost Government a staggering $450 million each year (Ibid.). Apparently, too, noncompetitive procurement policies have burgeoned rather than diminished. The General Accounting Office informed Congress in 1978 that only about 7 percent of Defense Department procurement dollars were awarded through formal advertising, down from 11 percent in 1971 (Statement of Walton H. Sheley, Jr., Deputy Director Procurement and Systems Acquisition, General Accounting Office,

Competitive Procurement, Hearing Before The Task Force On Government Inefficiency Of The Committee On The Budget, House of Representatives 96th Congress 1st Session 9 November 1979, USGPO 1980: 6).

Government agency lock-in to particular telematics suppliers— supported by both insufficient standardization efforts and extensive noncompetitive or sole-source procurements—bolsters the *entire* U.S. data processing industry by effectively guaranteeing companies selected shares of the Government market. Perhaps the Government's most vital role as a market force—although this is true mainly for data processing and not for telecommunications—has been thus to cushion growth for many companies. Although IBM's overall market share in the computer industry has hovered between 65-75 since the late 1950s (Brock, 1975: 21), within the Government market it has achieved a comparable measure of market power in not one of the five major computer submarkets (see Table 14). Moreover, in several submarkets IBM is subject to effective competition by other companies—Univac (Sperry) and Control Data particularly. In all likelihood this does not chiefly imply head-to-head competition for specific contracts. "Sole-sourcing"—noncompetitive procurements—ensure that the latter takes place but infrequently. Government, rather, seems to be force-fed a diet supplied by the entire computer industry. Value figures for computer expenditures by each agency and department are not accessible; but even data on the *number* of computers owned or leased by each agency reveal a spread (Table 15) (Digital Equipment Corp. and Data General appear most often in a top rank because they produce small, less expensive computers).

But acquisition of computers and components is only part of a broader picture. Contract services, including timesharing, systems analysis and design and programming services, maintenance and operating services, and other associated categories of service, account for a growing segment of total Government operating expenditures on data processing. As Lester Fettig of the Office of Federal Procurement Policy at the Office of Management and Budget told a Senate Subcommittee (Statement of Lester Fettig, *Contracting Out Of Defense Functions*

TABLE 14
1979 IBM Market Share of Selected Government Computer Submarkets (Percentage)

Submarket	IBM Share (%)	Closest Rival	2d Closest Rival
Central Processing Units	28	19 (CDC)	11 (Univac)
Input/Output	31	10 (CDC)	10 (Univac)
Storage Units	24	16 (Univac)	13 (CDC)
Communications Terminals	12	12 (Univac)	11 (Burroughs, Honeywell)
Other Components	45	12 (Univac)	4 (CDC)
Total	27	14 (CDC)	13 (Univac)

Source: "Automatic Data Processing Activities in the United States Government as of the and of Fiscal Year 1979," Summary, General Services Administration Automated Data and Telecommunications Service, September 1980, Washington, DC, USGPO, National Stock Number 7610-00-111-8245: 60-65.

TABLE 15
Rank-Order of Computer Manufacturers by Number of Computers Per Agency*

Agency/Mfgr.	BUR	CDC	DEQ	DGC	HON	HPC	IBM	MOD	UNI	XER	Total #
Agriculture			3	1		3	2				143
Commerce			2	1					3		394
Energy			1	2	3						3390
EPA			1	2	3						158
HEW			1				2		3		387
Interior			1	3	2						141
NASA			1					2	3		1862
Treasury					1		2		3		209
TVA			3	1	2	3					119
Transportation			3			1			2		373
VA			1		3	2					286
Other Civilian				2		3			1		373
Total Civilian	10	8	1	2	7	6	4	5	3	9	7835
Air Force			2		3				1		2393
Army			3			1			2		1579
Navy			2		3				1		2120
Other DoD			3	1				2			406
Total Dod	8	6	2	7	5	4	3	10	1	9	6498

Source: "Automatic Data Processing Activities in the United States Government as of the end of Fiscal Year 1979," Summary, GeneralServices Administration Automated Data and Telecommunications Service, September 1980, Washington, D.C. USGPO, National Stock Number 7610–00–111–8245: 36–53.
*Note: Table 4 charts the number of computers by agency, and not the value by agency, and hence overestimates the importance of small computer manufacturers like Digital Equipment and Data General relative to large mainframe suppliers like Control Data and IBM. BUR=Burroughs; CDC=Control Data, DEQ=Digital Equipment; DGC=Data General; HON=Honeywell; HPC=Hewlett-Packard; IBM=International Business Machines; MOD=Modular Systems; UNI=Sperry Rand.

And Services, Hearings Before The Subcommittee On Manpower and Personnel Of The Committee On Armed Services, United States Senate, 95th Congress 1st Session 12 July 1977, USGPO 1977: 38):

> When the Government initially began to use automatic data processing, there was no viable commercial ADP services market to supply its needs. Consequently, most agencies developed a capability to provide their own services buying equipment and employing pesonnel. For more than ten years, there has been a vigorous and growing commercial ADP capability that could provide many of the services needed by Government, but little effort by agencies to review their existing activities or proposed "new starts" to determine if they are justified as exceptions to the policy of relying on the private sector.

This last-mentioned policy was developed and formalized through the Office of Management and Budget over decades—ultimately, in fact, formalized reliance on the private sector returns to the Civil War era—through various edicts. Its "general thrust," according to erstwhile Comptroller General Elmer B. Staats, "is that executive agencies will not start or continue an existing commercial or industrial activity to provide a product or service for their own use if such a product or service can be procured through the private enterprise system" (Ibid.: 4). Exceptions—as for military preparedness—may be made.

Table 16 delimits Federal department and agency ADP contract services as a percentage of total ADP operating and capital costs for 1979. It shows that over one quarter of public-sector functions currently making use of General Management Classification computer systems are operationally controlled by private contractors. Some agencies contract out half to two-thirds of total ADP operations. This trend is especially pronounced for military and military-related departments.

NASA, for instance, which farms over 58 percent of its ADP needs out to private services firms, has been criticized by a high-level workshop on space program computer capabilities for relying "too heavily on the recommendations and data processing suggestions of outside contractors when it should be doing most of the nitty-gritty decision making itself. And because of this dependence, the space agency's overall computer skills and abilities may be 10 to 15 years behind . . ." (Scannell, "Seminar: Nasa Should Develop Own Expertise," 1981: 22). The De-

TABLE 16
Contract Services as a Percentage of Total ADP Operating and Capital Costs by Government Agency 1979

Agency	Contract Services ($000s)	Total Operating & Capital Costs ($000s)	Percentage Contract Services
Agriculture	11,305	89,616	12.6
Commerce	27,533	125,623	21.9
Air Force	167,971	753,158	22.3
Army	135,012	820,365	16.4
Navy	95,179	539,765	17.6
Energy	191,315	287,391	66.6
EPA	30,639	49,058	62.5
FCC	1,415	5,384	26.3
FTC	2,305	3,536	65.2
GSA	83,977	130,998	64.1
HEW	209,074	433,423	48.2
Labor	19,455	47,890	40.6
NASA	128,078	220,122	58.2
NRC	2,353	4,601	51.1
SEC	282	2,279	12.4
State	4,312	19,432	22.2
Treasury	23,923	555,836	4.3
Total (includes other agencies)	1,296,245	4,764,207	27.2

Note: Does not include Special Management Classification Computer System costs.
Source: "Automatic Data Processing Activities Summary in the United States Government as of the end of fiscal year 1979," General Services Administration Automated Data and Telecommunications Service, September 1980, Washington, D.C., USGPO Stock No. 7610–00–111–8245: 94–98.

partment of Energy, which contracts for two-thirds of total computer activity, has also come under fire. In April 1976, for example, its Energy Research and Development Administration hired a private firm, Energy and Environmental Analysis, to "review and evaluate the adequacy" of its budget allocations and define its "information needs . . .and information system requirements" (Statement of Daniel Guttman, *Federal Government's Use of Consultant Services*, Hearings Before The Subcommittee On Civil Services And General Services Of The Committee On Governmental Affairs, U.S. Senate 96th Congress 1st Session 12 October 1979 USGPO 1980: 11–12). In the words of attorney Daniel Guttman the latter confers upon this private firm power to "tell our employees what they need to know" (Ibid.). Through such arrangements, institutionalized in a multi-billion dollar contractor-consultant bureaucracy, the Federal Government, according to another expert on these issues, "has, over the years, lost the expertise to perform key functions in-house and, perhaps even worse, has lost the capability to monitor or evaluate the consultants' work" (Ibid.: Statement of John Hanrahan, 90). Interviews with private sector managers reveal that they believe that a prominent drawback of relying upon commercial computer services firms is that "Management tends to lose control over part or all of its operations" ("Shifting The Government's Automatic Data Processing Requirements To The Private Sector: Further Study And Better Guidance Needed," Report By The U.S. General Accounting Office, FGMSD-78-22, 11 April 1978: 7). Are public sector managers for some reason different?

Commercial provision of computer services to Government confers substantive control over Federal functions upon firms responsive first and foremost to private interests. The Office of Management and Budget nonetheless deliberately decrees that "direct procurement of services, with all the associated functions being performed by the private sector, is the preferred alternative for meeting data processing requirements" (Ibid.:22).

Government's ability to administer and use information resources is in fact undergoing a far more general process of transformation, paralleling and responding to changes in the private sector. Let us turn to a vividly illuminating instance of these broader and deeper trends—the Paperwork Reduction Act of 1980.

"PAPERWORK REDUCTION" AND INFORMATION MANAGEMENT: GOVERNMENT AS TELEMATICS USER

With the submission of an "information collection budget" for 1981 by the U.S. Office of Management and Budget we are for the first time to have an extensive measure of control over the nature and extent of Federal paperwork. Mandated by Executive Order, this paperwork budget was codified quietly into law in November 1980, when Congress passed the Paperwork Reduction Act of 1980—which went into effect 1 April 1981. H.R. 6410—"To Reduce Paperwork And Enhance The Economy And Efficiency Of Government And The Private Sector By Improving Federal Information Policy-Making, And For Other Purposes"— appears at first as a benign and long overdue attempt to check a flood of Federal red tape. In reality it is a radical move to centralize oversight and control of Federal information activities in the executive branch, with profound consequences for the substance of Government regulatory activities.

"Paperwork reduction," to begin with, is a deliberate misnomer. According to Frank J. Carr, Commissioner of Automated Data and Telecommunications Service at the General Services Administration, "the act also might have been termed the Information Management Act of 1980" (*Paperwork Reduction Act Of 1980*, Hearings Before a Subcommittee Of The Committee On Government Operations, House of Representatives 96th Congress 2d Session on H.R. 6410, 7, 21, 26 February 1980, USGPO: 132). To this claim, New York Representative Frank Horton—a motive force behind the bill—responded (Ibid.): "You and I know what information management means, but the general public does not know . . . .I know what you are trying to say to us, but paperwork reduction is something people can relate to very quickly."

It is information management indeed, however, which forms the core of the new law. The Office of Management and Budget argued for the bill (Ibid.: 196) on the grounds that "its concept of integrated, life-cycle information management has a significant potential for cost savings. Improving and strengthening all facets of information management, including the collection, storage, utilization, manipulation, transmission, and dissemination of Federal information, is the principle (sic) goal." The Government should look, as Horton preferred, to "a new management discipline, that of managing information or paperwork as a resource like other resources such as money, personnel, and property, which have values and costs and are used in achieving program goals" (*Privacy And Confidentiality Report And Final Recommendations Of The Commission On Federal Paperwork*, Hearings Before A Subcommittee Of The Committee On Government Operations, U.S. House Of Representatives 95th Congress 1st Session 17 October 1977, USGPO: 11).

Information management means treating information as a commodity. This is hammered home repeatedly through congressional hearings: "(F)or over 100 years the Government has collected and managed information as though it were a free good," lamented Wayne Granquist, then Associate Director for Management and Regulatory Policy at OMB (*Paperwork Reduction Act Of 1980*: 89). "We can no longer afford to have Government look at data as free and have information and paperwork costs buried in overhead accounts where costs cannot even be identified, much less recovered," emphasized Horton (Ibid.: 33). Granquist brightened (Ibid.: 89): "The time has come when we must recognize that there are costs associated with information—both to the provider and the user . . . ."

Federal information collection and use is said to contribute both to higher taxes and sprawling bureaucracy—but *also* to direct and mounting costs for the *private sector providers* of information. Horton estimated that "over $100 billion a year is spent in and out of Government on Federal Information needs and handling" (Ibid.:33)—but how this staggering figure was derived is not made clear. Unraveled, the logic of paperwork reduction runs like this (*Privacy And Confidentiality Report*: 6). "Just as government has a responsibility to manage its own costs and resources efficiently and effectively, it also has an obligation to the private sector to minimize the burdens imposed by the government. To fulfill this obligation, we must consider the full costs of information requests, reporting and record-keeping, not merely the cost to the government."

Before turning to examine further this startling sea-change in Government information handling and management, it should be noted that previous legislation to guide Government in its information gathering dates to 1942—and the Federal Reports Act passed in that year. Paralleling this wartime public-sector initiative, a Business Advisory Council On Federal Reports was also founded in 1942. Currently (1979) its board of directors harbors representatives of Chrysler, the American Paper Institute, National Association of Broadcasters, Kodak, Sears, U.S. Steel, Dupont, National Association of Manufacturers, BF Goodrich, and the U.S. Chamber of Commerce. Financed by business, this Advisory Council works with the Office of Management and Budget and the General Accounting Office, as well as with individual departments and agencies, to review forms and Federal information collection programs, and to "provide knowledgeable input in the review process" from the business sector (Statement of the Business Advisory Council On Federal Reports, *Paperwork And Redtape Reduction Act Of 1979*, Hearings Before The Subcommittee On Federal Spending Practices And Open Government Of The Committee On Gvoernmental Affairs, U.S. Senate, 96th Congress 1st Session On S.1411, 1 November 1979, USGPO: 178). In addition to helping to "improve reporting forms" and to "simplify record keeping requirements" and "eliminate or consolidate data requests," the Advisory Council tries" (t)o monitor how government uses the information being collected" (Ibid.: 179).

Today, however, business is pushing for dramatic changes in Federal information collection, management and use. Briefly, with integration of telematics into production, distribution and administration functions of geographically dispersed companies, identification and rationalization of an isolable set of information and information-related costs becomes possible for the first time. Electronic "paperless processing" of data not only provides new means to measure and stimulate work flow but may also permit centralization of varied information management functions. An American Express advertisement, for instance, emphasizes that by relying upon a single credit card account for all travel and entertainment spending by employees, companies can achieve consolidated billing and related economies—"And don't forget the money you've saved slashing paperwork"

(American Express, 1981: 40–41). Coupled with a burgeoning commercial computer services market whose axial principle is value-of-service pricing, and proliferation of telematics equipment and services within Government itself, Government's own information management practices are not surprisingly coming in for intensive reevaluation. Indeed, as the largest employer of information—and of information workers—in the country, the Government is being asked to "take a leadership position in order to accelerate the way it makes it (sic) professional and administrative information workers more productive"—as a vice president of Xerox's Information Products Group puts it (Statement of Paul A. Strassman, *Communications Research And Development,* Hearings Before The Subcommittee On Transportation, Aviation And Communications, And The Subcommittee On Science, Research and Technology Of The Committee On Science And Technology, U.S. House of Representatives, 96th Congress 2d Session 20, 21, 22, 28 May 1980, USGPO: 72). As the perception that information is never free grows, in tandem with maturing telematics and information service markets and the application of market criteria to new realms, Washington—the information capital of the world—increasingly becomes an object of suspicion and even of open hostility for commercial information providers and users. For the latter, Federal information practices may be fraught with newly discoverable hidden costs, such as those associated with compliance with Government information requests. For information suppliers, Government presents an actual threat, in that it may be competing with a private sector provider—and furnishing a potentially profitable service free or at a cut-rate price. "The growth of information activities in our society poses unique problems with the relationships between government and the private sector," delicately notes a task force of the National Commission on Libraries and Information Science (Kirchner, "Study: Split Public, Private Information Roles," 1981: 17). Paperwork reduction is an aspect of this encompassing process, wherein public-and private-sector relations are broadly reconfigured to take account of the application of commercial market criteria to information gathering, processing, reproduction and transmission.

Inevitably this process, and the concomitant attempt to isolate information costs, threatens to alter the very organization and concrete nature of Government, because information cannot in practice be separated, as Robert Bruce puts it, "from substantive policymaking. Information gathering is essential to formulating policy" (Robert R. Bruce, *Paperwork Reduction Act of 1980:* 319). A study researched and published by the General Accounting Office illuminates and clarifies this claim. In a table entitled "Top Five Reporting Requirements In Terms Of Total Burden Hours By Most Burdensome Agencies," the General Accounting Office reports that the Toxic Substances Control Act inflicts 600,000 paperwork hours on the private sector; that a form "Approval of PCB Disposal Site and Record of PCB Storage and Disposal" demands 80,000 hours; and that the Occupational Safety and Health Agency generates 320,000 hours from the pri-

single control point for Federal Information Management . . . . will force a much overdue reexamination of internal agency procedures, capabilities, and policies . . .''(Ibid.: 2, 4–5).

At one level, consonant with the concerted push by business to dispense with Government regulation altogether where it occasions added cost without directly assisting corporate objectives, the Paperwork Reduction Act has simply allowed administrators to eviscerate Government activities. ''You can reduce the burden by reducing the frequency with which information is sent to the Government,'' one administrator imperturbably suggests (*Paperwork Reduction Act Of 1980:* 104). ''You can reduce the size of the sample . . . . You can also reduce the amount of information on an individual form, or you could eliminate the form. We have used all four of these methods . . .

> One good example of reducing the size of the sample, or eliminating the form, is OSHA. It knocked out completely its reporting requirements for all employers with 10 employees or less. That eliminated 40,000 American businesses from OSHA forms. It reduced the burden substantially (Ibid.: 104–105).

Only harsh disregard for occupational safety and health, or redoubtable ignorance of the relation between statistical policy and social reality, can underlie this innovation. And, insofar as the information collection budget established by the OMB is concerned, when Congressman Levitas queried this administrator further (''You say you have a paperwork budget. What is it? What do you tell the EPA?'' (Ibid.: 106)), he received the following reply:

> Let us say EPA comes in and says they want to occupy 40 million hours of the public's time next year in filling out forms. We look at those requirements and we look at the forms. We can say that we disagree. We can make suggestions about doing it another way. It is the same as a budget process in general.
>
> We will negotiate those totals. Once the total is established the Administrator of EPA would be required to stay within that total. We are trying to force trade-offs down the line and require that information no longer be looked at as a free resource. We want to require an agency head who wants to collect the information about situation x, to analyze the trade-offs and realize he cannot get as much information about something else that he might be interested in. He has to make priority decisions. (Ibid.).

But by what criteria? And with what sort of public access and accountability? *As a condition of scarcity is imposed by transforming information into a commodity, the substance of Government policy is challenged and transformed.*

A few other examples show how the Act has performed thus far. Although one of its provisions expressly exempted independent regulatory agencies from being subjected to unlimited information-budgetary control by the Office of Manvate sector through its Occupational Injuries and Illnesses Survey ("Federal Paperwork: Its Impact On American Businesses," Report By The Comptroller General Of The United States, General Accounting Office GGD-79-4, 17 November 1978: 25–26). For such statistics to be jumbled together indifferently with analogous figures for Commerce Department Census of Manufacturers forms (470,000 hours) and Defense Department Personnel Security Clearance Change Requests (92,000 hours) signals the emergence of an entirely new policymaking calculus in which substantive social goals must vie with the purported costs of regulation and, especially, with the reputed costs of *compliance* with Government strictures. "Whenever we legislate, regulate, or otherwise require the public to bear information costs," states congressman Horton (*Privacy And Confidentiality Report*: 6), "we must know what they are so that decisions can be made which will equitably distribute the costs and benefits between the public and private sectors. Full accounting is necessary to manage these costs." Such a conclusion is fully in keeping with the recommendations of the President's Project on Information Technology and Government Reorganization, whose reports in 1978 and 1979 underlay the Paperwork Reduction Act and whose team leaders were drawn from data processing and information departments at CBS, General Foods, Lockheed, Coca Cola and Blue Cross.

The Paperwork Reduction Act grants the Office of Management and Budget broad powers of control over Government informational activities, centralizing responsibility in a new Office of Information and Regulatory Affairs. This Office will oversee Federal information management activities such as automatic data processing and telecommunications, paperwork clearance, records management, the Privacy Act and statistical policy. The Act also forces agencies to create a senior position for oversight and periodic review of information management activities; this, it is hoped, will streamline and support OMB-agency relations. The 1942 Federal Reports Act, which governed clearance of information collection requests by agencies, but which exempted some 80 percent of all such requests through various loopholes, has now been strengthened as a mechanism "to minimize the public reporting burden" (*Paperwork Reduction Act Of 1980* 96th Congress 2d Session, U.S. House of Representatives, Report No. 96-835, 19 March 1980: 2). (The only surviving loopholes relate to intelligence and defense.) The Act finally establishes a Federal Information Locator System—inconceivable without massive computerization of Government operations—which will house descriptions of all information requests made by agencies as well as any reports or analyses derived from such requests. The system will identify duplication in agency and department reporting and record-keeping requirements; locate existing information that may meet Government needs; and assist OMB in deciding which agency requests for information should be approved (Ibid.). Supposedly such changes will result in "uniformity and consistency in policies for acquisition and management of advanced information technology"; and, by establishing "a

agement and Budget, these same agencies have been singled out for further attention. A recent GAO report, thus, asserts that the Civil Aeronautics Board, Interstate Commerce Commission, Federal Energy Regulatory Commission, Federal Maritime Commission and Federal Communications Commission "must take additional steps to further reduce the accounting and financial reporting burden on industry" ("Independent Regulatory Agencies Can Reduce Paperwork Burden On Industry," Report To The Director, Office Of Management And Budget By The U.S. General Accounting Office, AFMD-81-70, 7 July 1981: 1). The Federal Communications Commission has already done so. A just-released notice states: "The Commission, by its Regulatory Review Working Group, is undertaking a much broader review of all rules and policies with the objective of eliminating those which no longer serve the public interest" (General Docket No. 81-706, "Federal Communications Commission's List of Rules to be Reviewed Pursuant to Section 610 of the Regulatory Flexibility Act During 1981–1982," Notice Released 30 October 1981: 2). Mandatory homage to "public interest" guidelines set forth in the 1934 Communications Act should not deter recognition that the FCC is substantively concerned with, among other things: "(1) the nature of the economic impact the rule(s) has (or have) on the commenting party," in inviting businesses and other parties to comment on such matters (Ibid: 1).

A report in the *Wall Street Journal* gives further indications of the impact of the Paperwork Reduction Act. Together with more general budget cuts, the Act may threaten Government's ability to produce accurate and timely economic data, charges Courtenay Slater, a former top Commerce Department economist (McGinley, 1981: 21, 31). Some Bureau of Labor Statistics indicators, including state and local unemployment rates, are now already being based on scantier information "and others may be cut entirely in the next round of budget-slashing" (Ibid.: 21). Over the past 30 years, Government's statistical policy group has apparently been cut from 70 people to fifteen; now, based at the Office of Management and Budget, the group, according to Courtenay, is playing "third fiddle to regulatory reform and paperwork reduction" (Ibid.: 31). In this case there is a barely concealed irony about the situation: intended to cut costs, the Paperwork Reduction Act, to the extent that it engenders inaccuracy in Government economic indicators, may actually increase expenditures. "Billions of dollars in Social Security payments and union wages are tied to the consumer price index, the most widely watched measure of inflation. Government officials make policy decisions based on the statistics. And businessmen have become increasingly dependent on the figures as a guide in investment and planning" (Ibid.: 21). As an official of the Bureau of Labor Statistics asserts, "in the long run it's cheaper to keep good statistics" (Ibid.).

The full implications of paperwork reduction are far from clear. It is evident, however, that at minimum the push for improved information management subjects Government to an unprecedented test: Do Government functions and activities impose "unnecessary" costs on the private sector? Are Federal information demands cost-justified? We found above that a substantial component of Government telematics and information activity is performed by contract with commercial computer services firms, so that competition with private-sector suppliers is deliberately minimized. With the Paperwork Reduction Act a closely related constraint emerges. Data that Government may need to perform a host of old and new functions in its oversight of the commonweal will now be identified, gathered, stored and used with an abacus in hand to tot up a series of *imposed* information costs. Leave aside any pragmatic or conceptual questions about the definition of information costs—and there are many—for the underlying issue in paperwork reduction is just this, that information is never free. This radical claim both emanates from and itself helps to accelerate the erosion of the public sector and concomitant acceptance of some traditional public sector functions by private companies making use of telematics. Privatization is perhaps most visible in the recent travails of the U.S. Postal Service.

MONOPOLY AND THE MAIL: GOVERNMENT AS PROSPECTIVE TELEMATICS SUPPLIER

The recent history of the U.S. Postal Service (USPS) illustrates tellingly that the overall role of Government has not been simply to allow telematics to develop freely—but, instead, to assist in forcefully and rapidly privatizing and commercializing telematics services. In stark contrast to its enthusiastic encouragement for private telematics growth, on one side, and for militarization of telematics services, on the other, Government has sought to postpone and prohibit incorporation of telematics in the hands of a quasi-public civilian entity, the Postal Service. In this respect, current attacks on the Postal Service are directly analogous to assaults on PTT ministries, chronicled in Part Two.

That the Postal Service *must* enter and participate "vigorously" in development of an electronic mail system if it is to survive is plain (Sorkin, 1980: 116). Electronic mail and message systems have already been broadly endorsed by the private sector. Companies like Satellite Business Systems and Graphnet offer a line of products to serve this field, while others wait in the wings; predictions stress that private electronic mail will expand rapidly through the current decade. A General Accounting Office study, summarizing some of these, concludes that "(t)he expansion of electronic mail services will not wait for a Postal Service role to be determined nor does its expansion depend on the Service even having a role" ("Implications Of Electronic Mail For The Postal Service's Work Force," Report By The Comptroller General Of The United States, General Accounting Office, GGD-81-30, 6 February 1981: 17). A survey cited by Walter T. Marable, executive director of the USPS Research and Development Laboratories, estimates that some 350 of the top 500 Fortune companies will implement a form of electronic mail by 1982 ("Questions and Answers From Walter T. Marable, "*Communica-*

tions Research And Development, Hearings Before The Subcommittee on Transportation, Aviation, And Communications, And The Subcommittee On Science, Research, And Technology Of The Committee On Science and Technology, U.S. House of Representatives 96th Congress 2d Session, 20–22 May 1980: 472).

Electronic funds transfer systems—payments mechanisms in which processing and communications needed to effect economic exchange are at least partly dependent on electronics—are also mentioned frequently as likely to cause "significant losses of Postal Service volume, particularly in the business transactions component of first-class mail" (Sorkin 1980: 113). Treasury Department policies, responsive to private banking interests, have themselves contributed important additional force in this regard, by encouraging automatic direct deposit of social security checks, supplemental income payments, civil service retirement payments, and revenue sharing payments to city and state government (Ibid.: 115). Nearly 13 million monthly Treasury payments currently employ Direct Deposit/Electronic Funds Transfer capabilities ("Implications Of Electronic Mail . . .": 33). Estimates vary as to how much first class mail will be diverted annually as a result of electronic systems; the Commission on Postal Service claims that, by 1985, fully 23 percent of total first class mail volume (17.5 billion out of 73.5 billion pieces) will be lost (Ibid.: 115–116).

Meanwhile, second class service, since 1970, has seen dramatic price increases along with rate hikes for required transportation services. Faced with escalating postage costs and slower delivery, periodicals and newspapers (the most prominent contributors to second class mail service) have generated substantial interest in private mail distribution ("Implications Of Electronic Mail . . .": 33). *Better Homes and Gardens* relies upon private carriers in sixteen cities to deliver 500,000 copies—7 percent of total subscriptions (Sorkin 1980: 139–140, citing David McClintick, "Spurred by Rise in Postal Rates, Publishers Expand Use of Private Delivery Services." *Wall Street Journal* 13 August 1978: 42). *Time* likewise employs private firms to deliver two percent of its subscription circulation— 100,000 copies—and has added *Sports Illustrated* to its alternative delivery "experiments" (Ibid.). The most extensive private delivery system currently in use by a U.S. publication is that of the *Wall Street Journal*, which sends 15 percent of its 1.2 million weekday subscribers copies delivered outside the postal system (Ibid.). "(I)f postal rates continue to increase," according to Sorkin (Ibid.), "60 percent or more of magazine subscription circulation could be distributed by private firms within several years." The *New York Times, Newsweek* and Doubleday's Book Club Division all are also utilizing private delivery systems on an experimental basis (Ibid.: 140–141).

Alternative delivery systems had their earliest impact on fourth class mail. Parcel post, a key subcategory in this class, has been virtually usurped by a private firm—United Parcel Service. The Postal Service in 1970 delivered 800 million packages to United Parcel's 500 million; by 1979 Government carriers delivered only 200 million, while United Parcel transported fully 1.4 billion packages

("Implications of Electronic Mail . . ."': 33). Other private firms then launched express mail services, guaranteeing same-day or next-day delivery of packages; since commencing operations in 1973, one of these, Federal Express, saw revenues shoot skyward to $589.5 million by 1981 (Federal Express Corporation Annual Report 1981: 1). Claiming grandly that it "is nothing less than the logistics system of the information-based phase of the industrial revolution," (Ibid.: 5), Federal Express is dedicated to providing a business-oriented service able to match the velocity of a telematics-based economy: "The time taken to deliver a spare part necessary to return a machine to production, to provide a piece of diagnostic equipment to a hospital, to send a contract to a closing, or to speed a set of architectural plans to a job site, is crucial" (Ibid.). As fourth class USPS revenues decline, meanwhile, according to the Assistant Postmaster General for Customer Services, alternative delivery is a more immediate threat to the USPS than any of the new communications technologies, owing to its decisive impact on the total rate structure ("Implications Of Electronic Mail . . .: 33").

Other threats to the USPS again return to private telematics capabilities. Use by customers of the telephone to pay bills, a service increasingly touted by banks eager to automate payments functions further, cuts into first class mailings. And, at the other end, businesses are displaying considerable interest in use of toll-free "800" (WATS) telephone numbers which reduce the volume of potential reply mail (Ibid.).

What has the Postal Service done to protect its traditional turf? First, it has (belatedly) attempted to implement a new service called "E-COM"—Electronic Computer Originated Mail. E-COM is intended for the convenience of volume mailers to generate mail from data stored in electronic form; it will be limited to mailers who can generate at least 200 messages per transmission per Serving Post Office (SPOs) (25 Serving Post Offices around the nation have been designated; service began on 4 January 1982).

> The Postal Service will accumulate electronically received messages in its own computers, sort the messages in ZIP Code sequence for efficient delivery, print the messages on paper, and insert them into specially marked envelopes. Messages will then be delivered to addresses as First-Class Mail. Delivery areas near the SPOs will generally receive next business day service, while those more distant will be served the second business day ("Postal Service," *Federal Register* Vol. 46. No. 199, 15 October 1981, Notices: 50875)*

E-COM thus is of use to large-volume mailers—mass consumer advertisers mounting campaigns especially—and offers them the chance to avoid handling

---

*I should like to acknowledge David Bradbury for this reference.

hard copy all the way from computerized mailing list to ultimate message transmission.

That is, with one—vital—exception. When the Postal Rate Commission authorized E-COM as an experimental service in 1979, USPS was told it might furnish end-to-end electronic service, including telecommunications, *only* in the presence of a "demonstrated need" (Hirsch, "Controversy Over Ecom Nears Boiling Point," 1981: 7). The Postal Service was permitted to supply electronic mail services subject to the proviso that it would purchase needed transmission capacity from *private* carriers, rather than build its own network; moreover, prospective users of E-COM must make their own arrangements with private carriers to transmit data to participating Serving Post Offices. This proviso places an artificial limitation on the efficiency of the E-COM offering by bifurcating a potentially integrated end-to-end service into one segment, to be provided by private carriers, and another, to be furnished by the Postal Service. Even so, and although Postmaster General William F. Bolger has repeatedly stated that USPS has no intention of acquiring its own transmission facilities, commercial electronic mail vendors—both present and future—remain skeptical (Hirsch, "U.S. Postal Service Ecom May Be in Jeopardy," 1981: 6). Firms like Graphnet, AT&T, IBM and ITT, Sorkin (1980:127) writes, now "are putting tremendous pressure on the administration to limit the Postal Service's entry into electronic mail."

Their efforts have not gone unrewarded. In a joint filing by the Department of Justice and the National Telecommunications and Information Administration during the summer of 1981, the Reagan administration asserted that E-COM is unnecessary because commercial suppliers are already offering similar services and that there is a "significant chance" E-COM will be subsidized by taxpayers, to the discomfiture of E-COM's prospective competitors (Hirsch, "U.S. Postal Service Ecom May Be in Jeopardy," 1981: 6).

More recently, as sparring between the Postal Service and private suppliers of electronic mail services has continued—the latter charging, indeed, that USPS has rigged the timing and technical specifications of its E-COM offering to discourage participation by communications carriers, thereby showing a "demonstrated need" to take over the communications function itself—a broader legislative initiative has been introduced. H.R. 4758, introduced by Congressman Glenn English, is designed to ensure that the Federal Government does not compete with private providers of telecommunications and information services. The bill would prohibit Federal agencies from providing such services to anyone apart from other agencies, except when no practical alternative exists or when the service is in the national interest. It has broad relevance to the process of privatization detailed above, but it is specifically directed against any possibility that the Postal Service might provide telecommunications service as part of an end-to-end electronic mail offering. To foreclose such an eventuality, the bill forbids USPS from supplying telecommunications services without prior congressional approval (Kirchner, "Bill Would Curb Federal DP, Telecommunications," 1981: 11). Movement toward even a relatively small measure of competitive public or Governmental participation in telematics has thus been fiercely contested.

With or without passage of such legislation (which would have much more far-reaching impact), regulatory agencies have also been effective partners in what the Postal Service's Walter T. Marable characterizes as a "consistent effort to prevent the Postal Service from participating actively in the development and use of electronic mail technology" (*Communications Research And Development* 1980: 472). A General Accounting Office report observes that the ensconced regulatory framework—which subjects USPS to the Postal Rate Commission and to the Federal Communications Commission without making clear where their respective jurisdictional domains begin and end—"slows" USPS participation in electronic mail ("Implications Of Electronic Mail . . .": 8).

Federal Communications Commission inaction has been of especial significance to delaying provision by USPS of a second electronic service offering, INTELPOST—for International Electronic Post. INTELPOST would furnish customers with a rapid means of transmitting and receiving facsimile copies of letters, documents and graphics, by way of a digital facsimile network between the United States and numerous other countries' PTT authorities. The network would encompass Argentina, Belgium, Canada, France, West Germany, the Netherlands, Switzerland and the United Kingdom. Two INTELPOST transmission facilities were constructed within the United States, at New York and at Washington, D.C.; field trial contracts were awarded to two international record carriers. However, in November 1979 and January 1980, the FCC twice rejected tariffs proposed by these carriers for the INTELPOST offering on grounds that the tariffs comprised unlawfully discriminatory abrogation of the resale and shared use restrictions governing employment of international circuits. Tariff rejection occurred in spite of the Commission's patent and simultaneous interest in *promoting* liberalized resale and shared use provisions for *private* users of international circuits, as chronicled in Part Two. The FCC's charge that INTELPOST tariffs were unlawfully discriminatory, moreover, was made despite the willingness of foreign PTT administrations, in contradiction to their policies regarding private firms, to authorize international resale in conjunction with U.S. Postal Service tariff offerings. FCC intransigence meant simply that *only* private companies should be allowed to avail themselves of liberalized resale and shared use provisions. INTELPOST was forced to begin service indirectly, through Canada, and was unable to lease circuits directly from PTTs as a result of FCC policies taken at the U.S. end.

One potentially important tactic for the Postal Service in combating at least some of these obstacles stems from the private express statutes, which basically guarantee that Government shall have a legal monopoly over first class mail (Sorkin 1980: 133). Private express statutes have already been modified to allow private companies to deliver parcels, packages and time-sensitive items requiring rush delivery; thus, "extremely urgent" letters—defined as such if the amount paid for private carriage is more than three dollars or twice the applicable U.S. first-class postage rate, whichever is greater—may be delivered by Federal Express or other competing carriers. In 1979, then President Carter declared that the private express statutes should not be extended to cover electronic message transmission (Ibid.: 129). On 2 November 1981 legislation was introduced into the United States Senate to permit the carriage of mail, including first class mail, by private express or carrier, thereby effectively overturning the private express statutes altogether (*Private Mail Carriage Act Of 1981,* 97th Congress 1st Session, Draft).

It is not "technology" that is being freed to develop as rapidly as possible, but technology in private hands. Ironically, there is a hint that the Postal Service has bound itself at times willingly to these harsh strictures—first by agreeing not to engage in end-to-end electronic mail service and, more recently, by declaring it would "not hesitate to suspend the Private Express Statutes" on its own if the postal unions walked out on strike (Postmaster General William F. Bolger as quoted in Hirsch, "Intelpost Expansion Expected In Two Months," 1981: 19). Postal unions, concerned that in the face of trends outlined above, the Postal Service's response appears to stress massive automation and concomitant layoffs, are thus informed that, should they choose to fall back on the only bargaining weapon they possess, Postal Service administrators are fully prepared to commit organizational harakiri. Thus the prospective fortunes of the Postal Service perfectly match business' evident broader objective of "turning over government responsibilities to the private sector," while "dismantling the public sector" (Lee, 1981: 30).

The three themes of this Part round out our appraisal of telematics and Government. As a telematics *market* Government is farming out vast tracts of its public duty to private firms, helping hugely to sustain them in the process. As a telematics *user* Government has been subjected to the radical notion that information is a commodity to be measured and budgeted and controlled in support of program goals—with corresponding shifts in the substance and character of its policymaking functions. Finally, as a telematics *supplier* Government is being shown the door, at least insofar as activities that compete with the private sector are concerned. That the range and number of these activities are expanding is due to the aggressive incorporation of telematics by the private sector and not to any sudden enthusiasm for competition voiced by Government. Government continues, however, as shown in Parts One and Two, to broaden and deepen the hold of private enterprise over telematics technology and services, in search of unparalleled centralization of world markets. For these reasons the role of Government in the computerization of society must be considered of cardinal importance.

BIBLIOGRAPHY

Advertising Age. " 'Demassification': Coping with Splinters," 9 November 1981: 48.
Ahern, Veronica M. "The Introduction of New Services in the International Marketplace," *Telecommunication Journal* Vol. 47 No. VI (1980): 359–360.
American Enterprise Institute. *Telecommunications Law Reform*. Legislative Analysis No. 12, 96th Congress. Washington, DC: American Enterprise Institute, February, 1980.
American Express Company. Advertisement, *United Mainliner* Vol. 25 No. 1 (1981): 40–41.
American Telephone & Telegraph Company. *Annual Report 1980*.
"At the Office with Howard Anderson: The View From The Sidelines," *Computerworld* 29 June 1981: 3.
"AT&T Accord Clouds Outlook On Phone Bonds." *Wall Street Journal* January 1982: 8.
"AT&T Rate Rise On Four Services Set For Thursday," *Wall Street Journal* 11 May 1981: 7.
"AT&T Says It May Look Abroad to Offer Some Debt and Build Production Plants." *Wall Street Journal,* 20 January 1982: 8.
"AT&T Share Owners Newsletter." January 1982.
"AT&T To Be Freed From Licensing Bell Labs Patents," *Wall Street Journal* 29 January 1982: 43.
"Automated System Speeds Operations For International Bank's U.S. Branch," *Communications News* July 1980: 38.
Bargellini, Pier L. "Commercial U.S. Satellites," *IEEE Spectrum* October 1979: 35.
Barna, Becky. "Bucking The System," *Datamation* July 1980: 50–52.
"Bell 'a Withering Corporation' Under Consent Decree, Attorney Says," *Computerworld* 18 January 1982: 11.
Block, Victor. "AT&T Files WATS Revision, Eyes Private Line Boost," *Telephony* 29 September 1980: 12.
Block, Victor. "Congress Fears Consent Decree Will Lead to Higher Local Rates," *Telephony* 8 February 1982: 11–18.
Block, Victor. "FCC Ends Ban On MTS, WATS Resale, Shared Use," *Telephony* 5 November 1981: 12.
Block, Victor, "1980—A Year of Major Changes," *Telephony* 12 January 1981: 66.
Blumenthal, Marcia, "Industry's First Concern: Local Transmission," *Computerworld* 18 January 1982: 10.
Blumenthal, Marcia. "Processing Services A Whole New Ball Game," *Computerworld* 1 December 1980: 1, 10.
Blumenthal, Marcia. "Processing Services Expected to Grow 17%/year," *Computerworld* 1 December 1980: 12.
Blumenthal, Marcia. "Revenue Up 21% in 1980 For Computer Services," *Computerworld* 6 July 1981: 73–75.
Bolter, Walter G., and David A. Irwin. "Depreciation Reform—A Crucial Step In Transforming Telecommunications To A Free Market." Washington, DC: September 1980.
Brady, Robert A. *Business As A System Of Power*. New York: Columbia University Press, 1943.
"Britain Bids to Breach Europe's Telecommonopolies," *The Economist* 27 June 1981: 71–72.
Brock, Gerald. *The Telecommunications Industry*. Cambridge, MA: Harvard University Press, 1981.
Brock, Gerald. *The U.S. Computer Industry*. Boston, MA: Ballinger, 1975.
Brown, Charles L., 1981. Statement Before The Committee On Commerce, Science, and Transportation, U.S. Senate, 97th Congress 1st Session, On S. 898, *Telecommunications Competition and Deregulation Act of 1981*, 2, 11, 15, 16, 19 June, 1981, Serial No. 97-61. Washington, DC: USGPO, 406–431.
Brown, Charles L. Statement Before The Committee On Commerce, Science And Transportation, U.S. Senate, 97th Congress 1st Session, 25 January 1982.

Brown, Merrill. "Bill Shifts Phone Role to Military," *Washington Post* 27 September 1981: A1.
Bruce, Robert R. Statement Before The Subcommittee On Information And Individual Rights Committee On Government Operations, U.S. House of Representatives, Hearings On H.R. 1957, *International Communications Reorganization Act of 1981* 97th Congress 1st Session, 31 March, 2 April 1981: 46.
Burkert, Herbert. "German PTT Meets Criticism from 'Monopoly Commission'," *Transnational Data Report* Vol. 4 No. 4 (1981): 9–10.
"Business Calls for End to PTT Monopoly," *Transnational Data Report* Vol. 3 No. 8 (1980): 16.
"Bypassing Ma Bell . . ." *Wall Street Journal*, 5 February 1982: 29.
"California Assails AT&T Accord, Says It Plans to Intervene," *Wall Street Journal* 15 January 1982: 12.
Chace, Susan. "Outside Suppliers Find That Ringing Up Sales With AT&T Isn't Easy," *Wall Street Journal* 29 January 1981: 1.
Chace, Susan and White, James A. "IBM to Reenter Services Market In Computer Area." *Wall Street Journal* 3 February 1982: 2.
Chamoux, Jean-Pierre. "International Telecommunications Policies: a European Prospective View," *Transnational Data Report* Vol. 4 No. 1 (1981): 22.
Chandler, Alfred D. *Strategy and Structure: Chapters in the History of the American Industrial Enterprise*. Cambridge: MIT Press, 1962.
Chandler, Alfred D., Jr. *The Visible Hand*. Cambridge, MA: The Belknap Press, 1977.
Chase, Mel. "AT&T International: A New Voyage Into The Competitive World Marketplace," *Bell Telephone Magazine* Vol. 60 No. 2 (1981): 11–13.
Chung, William K., and Gregory G. Fouch. "Foreign Direct Investment in the United States in 1979," U.S. Department of Commerce, Bureau of Economic Analysis, *Survey Of Current Business* Vol. 60 No. 8 (1980): 38–51.
"Coalition of Communications Users Groups Forms Telecause," *Communications News*, November 1981: 78–79.
Cochran, Thomas C. *200 Years of American Business*. New York: Basic Books, 1972.
Cole, Robert J. "Sears Will Purchase Dean Witter In Plan to Offer Financial Services," *New York Times* 9 October 1981: A1.
Communications Satellite Corporation. *Annual Report 1980*.
Communications Satellite Corporation, Office of Public Affairs. "Comsat Guide to the Intelsat, Marisat and Comstar Satellite Systems," 1981.
Computer Sciences Corporation. *10-K Report to the Securities and Exchange Commission*, 31 December 1979.
Control Data Corporation. *10-K Report to the Securities and Exchange Commission*, 31 December 1979.
Cornell, Nina W., Daniel Kelley and Peter R. Greenhalgh. "Social Objectives and Competition in Common Carrier Communications: Incompatible or Inseparable?" Federal Communications Commission Office of Plans and Policy Working Paper Series No. 1, April 1980.
Craig, L. C. "Office Automation at Texas Instruments, Inc." Chapter 10, pp. 202–214 in Mitchell L. Moss, ed., *Telecommunications and Productivity*. Reading, Mass.: Addison-Wesley 1981.
Crane, Rhonda J. *The Politics of International Standards: France and the Color TV War*. Norwood, NJ: Ablex 1979.
Criner, James G. "Telecommunications resale: A policy analysis," *Telecommunications Policy* 1 (1977): 319–328.
Darlington, Roger. "The American Telecommunications System," Post Office Engineering Union Report, April 1981.
"The Datamation 100." *Datamation* July 1980: 98–101.
"The Datamation 100." *Datamation* June 1981: 91–192.
deButts, John D. Statement, U.S. Congress, *Hearings On Competition In The Telecommunications Industry* (1977): 17.
"Defense Department Lists Top 100." *Aviation Week & Space Technology* July 1980: 64–67.

De Jonquieres, Guy. "British Aerospace and IBM Link Up over Europe Satellite Plans," *Financial Times* 24 June 1981: 1.
"Deregulation Roils the Telex Market." *Business Week* 22 December 1980: 66.
Digital Equipment Corporation. *10-K Report to the Securities and Exchange Commission,* 31 December 1980.
Directions Des Affaires Industrielles Et Internationales, Direction Generale Des Telecommunications. *Annual Report 1980.*
Donne, Michael. "Industry Welcomes MoD Contract," *Financial Times* 27 April 1981: Financial Times Survey Communications, II.
Dooley, Ann. "CAD/CAM Seen Reshaping U.S. Work Habits," *Computerworld* 20 April 1981: 25.
Du Boff, Richard B. "Business Demand and the Development of the Telegraph in the United States, 1844–1860," *Business History Review* Winter 1980: 459–479.
"Dutch Publishers Sponsor Report Attacking PTT." *Transnational Data Report* Vol. 3 No. 8 (1980): 21.
"The EC Nations May Turn on Each Other." *Business Week* 2 November 1981: 62.
EDP Industry Report. 28 May 1980 (Vol. 15 Nos. 23, 24).
Eger, John M. "The International Information War," *Computerworld Extra!* 18 March 1981: 103–119.
Electronic Data Systems. *10-K Report to the Securities and Exchange Commission,* 31 December 1979.
"Electronic Mail Cuts Bank's Phone Dependence." *Computerworld* 29 June 1981: 19.
Emmett, Ralph. "Citishare Or Citigrab?" *Datamation* March 1981: 47–48.
Fargo, Dan S. "Int'l telecom spending continues its climb," *Telephony* 23 February 1981: 51–61.
"An FCC Plan Backfires Abroad," *Business Week* 28 July 1980: 56.
Federal Express Corporation. *Annual Report 1981.*
"Federal Government Endorses Version Of X.25, Seeks Comments." *Data Channels* Vol. 7 No. 11 (1980): 4.
Ferris, Charles D. "Telecommunication Policies for the 80s Will Call for New Vitality, Equity, and Freedom," *Communications News* December 1980: 30.
"The 500." *Fortune* 4 May 1981: 322–349.
"The Forbes 500's." *Forbes* 11 May 1981: 258–312.
"Ford's Dial-Up Design Technique." *Telecommunications* August 1981: 46–48.
"Foreign Agency Income." *Advertising Age* 20 April 1981, Section 2: 1–48.
"Foreign Ownership in France." *Multinational Monitor* Vol. 2 No. 9 (1981): 15.
"Form Group for Persons Involved in Resale Transmission Business." *Communications News* March 1981: 15.
"French Government States DP Plans." *Transnational Data Report* Vol. 4 No. 7 (1981): 29.
General Telephone and Electronics Corporation. *10-K Report to the Securities and Exchange Commission,* 31 December 1979.
Gilchrist, Bruce and Milton R. Wessel. *Government Regulation of the Computer Industry.* New York: AFIPS Press 1972.
Goldstine, Herman H. *The Computer from Pascal to Von Neumann.* Princeton: Princeton University Press 1972.
Grad, Frank P., and Daniel C. Goldfarb. "Government Regulation Of International Telecommunications," *Columbia Journal of Transnational Law* 15 (1976): 384–472.
"GSA Eases CSC Suspense." *Datamation* March 1981: 96.
Herman, Edward S. *Corporate Control, Corporate Power.* A Twentieth Century Fund Study. Cambridge MA: Cambridge University Press 1981.
Hewlett-Packard Company. *10-K Report to the Securities and Exchange Commission,* 31 December 1979.
Hirsch, Phil. "Both Tymnet, SBS Planning To Build Wideband Nets," *Computerworld* 24 August 1981: 1.
Hirsch, Phil. "Controversy Over Ecom Nears Boiling Point," *Computerworld* 2 November 1981: 7.

Hirsch, Phil. "Intelpost Expansion Expected In Two Months," *Computerworld* 23 February 1981: 19.
Hirsch, Phil. "Judge Lets AT&T Use DoD Study," *Computerworld* 31 August 1981: 16.
Hirsch, Phil. "Major Industry Battle Brewing Over Wideband Dems Planned By Tymnet, SBS and Isacomm," *Computerworld* 7 September 1981: 57.
Hirsch, Phil. "Nationwide Data Net Planned by Isacomm," *Computerworld* 10 August 1981: 13.
Hirsch, Phil. "U.S. Postal Service Ecom May Be In Jeopardy," *Computerworld* 10 August 1981: 6.
Hirsch, Phil. "User Faults New Comms Act Rewrite," *Computerworld* 22 June 1981: 8.
Hirsch, Phil. "Western Union, IRCs Dispute Gateway Ruling," *Computerworld* 3 November 1980: 36.
Hoard, Bruce. "IBM to support X.25, X.21 in U.S.," *Computerworld* 3 August 1981: 1, 6.
Hoard, Bruce. "Traffic in Packet Networks Seen Doubling," *Computerworld* 8 June 1981: 13.
Hobsbawm, Eric. *The Age of Capital 1848–1875*. New York: Charles Scribner's Sons 1975.
Holsendolph, Ernest. "House Draft On AT&T Submitted," *New York Times* 11 December 1981: D1, D4.
Holsendolph, Ernest. "Panel Told of Phone Worries." *New York Times* 10 February 1982: D6.
Honeywell Corporation. *10-K Report to the Securities and Exchange Commission*, 31 December 1979.
"House Unit Clears Bill to Ease Curbs On Western Union," *Wall Street Journal* 23 October 1981: 10.
"How Some Typical Customers Use SDX® Services," *American Satellite Capabilities and Service Profile* (1981?).
"IBM and British Firm Discuss Venture To Offer Satellite For Business In Europe." *Wall Street Journal* 25 June 1981: 30.
"ICA President Lloyd Isaacs Discusses Bank and Association Goals," *Communications News* June 1981: 54.
"Incorporate Fax Users Association." *Communications News* March 1981: 20.
"Index: 118 Mergers, Acquisitions Occurred in '81 Services Industry," *Computerworld* 25 January 1982: 79.
"Information Service Tailors Cost Control System to Arco Unit's Individual Needs." *Communication News* July 1980: 28.
Inglis, Andrew F. "Satellite Availability," *Satellite Circuit* No. 2 1981: 4–5.
"International Accounting Firms Consolidate Worldwide." *Multinational Business* No. 3 1980: 1–13.
International Business Machines. *Annual Report 1979*.
"Investor Confidence In AT&T Debt Shrivels, U.S. Panel Is Warned," *Wall Street Journal* 10 February 1982: 39.
"Irish PTT Looks to Private Sector for Telecom Financing." *Telephony* 23 March 1981: 37–38.
"ITT: Groping For A New Strategy." *Business Week* 15 December 1980: 77.
Jansky, Donald. "The Use of Radiotelecommunications in the Federal Government of the United States," *Telecommunication Journal* Vol. 47 No. VI (1980): 314–320.
"Japan Intends to Go on the Offensive, Ask U.S. to Drop its Own Barriers to Trade," *Wall Street Journal* 5 November 1981: 33.
"Japan takes aim at IBM's World," *World Business Weekly* 20 April 1981: 30.
"Joint Venture Formed for UK Leased Service." *Communications News* May 1981: 7.
Kaplan, Gadi. "Three Systems Defined," *IEEE Spectrum* October 1979: 42.
Karten, Howard. "Competition Called Users Insurance," *Information World* May 1980: 24.
Kass, Elliott M. "Geller Tells Users U.S. Needs Telecom Bill," *Information Systems News* 19 May 1980: 80.
Keatley, Robert. "Japan and Europe: Also at Loggerheads on Trade," *Wall Street Journal* 11 May 1981: 27.
King, John Leslie, and Kenneth L. Kraemer. "Cost as a Social Impact of Information Technology," Chapter 4, 93–130 in Mitchell L. Moss, ed., *Telecommunications and Productivity*. Reading, MA: Addison-Wesley 1981.

BIBLIOGRAPHY

Kirchner, Jake. "Administration to Push Services Trade Overseas," *Computerworld* 11 May 1981: 84.
Kirchner, Jake. "Bad Tidings in Store for Federal Meet Attendees," *Computerworld* 23 February 1981: 15.
Kirchner, Jake. "Bill Would Curb Federal DP, Telecommunications," *Computerworld* 2 November 1981: 11.
Kirchner, Jake. "Federal CPU Inventory Up," *Computerworld* 25 May 1981: 67–68.
Kirchner, Jake. "GAO Hits Federal Outlays for Software Upkeep," *Computerworld* 23 March 1981: 13.
Kirchner, Jake. "Hearings Open On Bill Forming Info Institute; NTIA Negative," *Computerworld* 1 June 1981: 13.
Kirchner, Jake. "International Communications Bill Draws Fire," *Computerworld* 13 July 1981: 5.
Kirchner, Jake. "State Fights to Keep Hold on Communications," *Computerworld* 20 July 1981: 13.
Kirchner, Jake. "Study: Split Public, Private Information Roles," *Computerworld* 7 September 1981: 17.
Lee, Susan. "Privatization: Nobody Does It Better," *Wall Street Journal* 18 November 1981: 30.
Lewis, Paul. "Common Market Dreams Wear Thin," *New York Times* 8 November 1981: F2.
Lohr, Steve. "Antitrust: Big Business Breathes Easier," *New York Times* 15 February 1981: F1.
"Ma Bell Unbound." *Wall Street Journal,* 12 January 1982: 32.
MacAvoy, Paul W. "Deregulation: A Letter to George Bush," *New York Times* 30 August 1981: F3.
Malik, Rex. "ICL Spearheading On U.S. DP Status Abroad," *Computerworld* 23 June 1980: 25–26.
"Manufacturers Hanover Trust Integrated Telex/Private Net," *Communications News* December 1980: 61.
Martin, James. *Future Developments In Telecommunications.* Second Edition. Englewood Cliffs, NJ: Prentice-Hall 1977.
McGinley, Laurie. "Economic Data Could Be Hurt By Budget Cuts," *Wall Street Journal* 27 November 1981: 21, 31.
MCI Communications Corporation And MCI Telecommunications Corporation, In The United States Court of Appeals For The Seventh Circuit, Nos. 80-2171, 80-2288, MCI *v.* American Telephone And Telegraph Company, *Brief Of Appellees,* 12 March 1981.
Melody, William H. "Interconnection: Impact on Competition, Carriers and Regulation," 1260–1271 in U.S. Senate, Hearings Before The Subcommittee On Antitrust And Monopoly Of The Committee On The Judiciary, 93d Congress 1st Session On. S. 1167, *The Industrial Reorganization Act,* Part 2 The Communications Industry, 30, 31 July; 1, 2, August 1973. Washington, D.C.: USGPO 1973.
"Mike Woody and Dan Grove Tell of TCA Plans and Progress." *Communications News* September 1981: 52-57.
"Missing Computer Software." *Business Week* 1 September 1980: 46–54.
"Mitterand: Why Nationalization Will Work." *Wall Street Journal* 7 October 1981: 27.
"More Comments On The ATT-DOJ Settlement," *Telephony* 1 February 1982: 38.
"More Satellites Crowding the Skies Mean More Useful Business Services." *Communications News* March 1981: 52–53.
Morgan, Walter L. "The Next Decade," *Satellite Communications* January 1981: 20–29.
"The New Attack On Bigness Takes Shape," *Business Week* 12 November 1949: 26.
"News Faces in an Age of Megadeals." *Forbes* 6 July 1981: 83.
"New Glamour for the Office PBX." *Business Week* 13 April 1981: 122.
Nichols, Richard B. "International Facilities Planning," *Telecommunication Journal* Vol. 47 No. VI (1980): 361.
"Nippon Telegraph Is Buying Abroad." *New York Times* 5 September 1981: 27.
Nora, Simon, and Alain Minc. *The Computerization of Society.* Cambridge: MIT Press 1980 (first published as *L'Informatisation de la société.* Paris: La Documentation Française 1978.)
"The 100 Largest U.S. Multinationals." *Forbes* 6 July 1981: 92–94.
O'Neill, Paul. "Telecommunications Users Groups," *Telematics* May 1981: 56–59.

Oettinger, Anthony. "Information Resources: Knowledge and Power in the 21st Century," *Science* 4 July 1980: 191–198.
"Organize Group for Iowa Users of Telecom Services." *Communications News* June 1981: 21.
"P&G Heads Drive for More Europe TV Time." *Advertising Age* 19 January 1981: 1, 102.
Peltu, Malcolm. "U.S. Hits Telecom Monopoly," *Datamation* January 1982: 80–88.
Pine, Art. "U.S., Common Market on Blowup Path," *Wall Street Journal* 18 November 1981: 35.
Pollack, Andrew. "Bell Upheld In Ruling On Growth," *New York Times* 5 September 1981: 25, 27.
Pred, Allan R. *Urban Growth and the Circulation of Information. The U.S. System of Cities, 1790–1840.* Cambridge, MA: Harvard University Press 1973.
Pred, Allan R. *Urban Growth and City-Systems in the United States, 1840–1860.* Cambridge, MA: Harvard University Press 1980.
"Progress, Many Problems Cloud AT&T-DOJ Settlement." *Telephony* 1 February 1982: 11–15.
"Report Calls for Telecom Free-For-All in the U.K." *Telephony* 25 May 1981: 49.
"The Resale Business in Phone Lines." *Business Week* 13 July 1981: 68.
"The Reciprocity Boomerang." *Wall Street Journal,* 4 February 1982: 30.
Rodgers, William. "Return to Sender," *Harper's,* December 1981: 22–27.
Rubiner, Betsy. "He'll Try to Keep Britain in Computers," *New York Times* 30 August 1981: F9.
"SBS, British Telecom Int'l Tell Joint Satcom Plans," *Telephony* 25 January 1982: 47.
"SBS Seeks Authority to Provide Service between U.S. and Canada." *Telephony* 9 February 1981: 22.
Scannell, Tim. " 'Fortune' Survey Finds No Loss of Corporate Control in Move to DDP," *Computerworld* 23 March 1981: 10–11.
Scannell, Tim. "Seminar: NASA Should Develop Own Expertise," *Computerworld* 10 August 1981: 22.
Schiller, Herbert I. *Mass Communications and American Empire.* Boston, MA: Beacon Press 1971.
Schultz, Brad. "ACS: Will It Be Baby Bell's Problem Child?" *Computerworld* 7 December 1981: 5.
Schultz, Brad. "Integrated Teletex Network Predicted for UK by '82," *Computerworld* 8 June 1981: 12.
Schultz, Brad. "Large IBM Users Planning On X.25 Now," *Computerworld* 24 November 1980: 1, 5.
Shayon, Robert Lewis. "Television International," Chapter 4, 41–55 in George Gerbner, ed., *Mass Media Policies in Changing Cultures.* New York: John Wiley & Sons 1977.
Sherman, Kenneth. "How the Common Carriers Line Up," *Computerworld* 29 June 1981: 4.
Sirbu, Marvin A., Jr. "The Innovation Process in Telecommunications," Chapter 9, 184–198 in Mitchell L. Moss, ed., *Telecommunications and Productivity.* Reading, MA: Addison-Wesley 1981.
Smith, Anthony. *The Geopolitics of Information.* New York: Oxford University Press 1980.
Sorkin, Alan L. *The Economics of the Postal System.* Lexington, MA: D.C. Heath 1980.
Sperry Rand Corporation. *10-K Report to the Security and Exchange Commission,* 31 December 1979.
Standard & Poor's Industry Surveys. *Office Equipment Systems and Services,* 15 May 1980.
Standard & Poor's Industry Surveys. *Office Equipment Systems and Services Current Analysis.* 26 June 1980.
"Stripping Ma Bell." *The Economist,* 16 January 1982: 13–15.
"Swiss Association of Telecommunications Users." *Telecommunications Policy* Vol. 5 No. 1 (1980): 41–43.
Tanenbaum, Morris. Statement Before Committee On The Judiciary, U.S. Senate 97th Congress 1st Session 25 January 1982.
"TCA Advocates Changes." *Communications News* September 1981: 57.
"The Technological Bubble." *Intermedia* January 1981: 14.
"Telecommunications Managers Association, UK." *Telecommunciations Policy* Vol. 5 No. 1 (1980): 40.

Thomas, Willard and Cinda Thomas. "Transponders: The Advantages of Purchase," *Educational and Industrial Television* Vol. 13 No. 5 (1981): 94–95.
Thompson, Geoffrey. "Challenging CEPT Agreements," *Telephony* 26 January 1981: 68–73.
"Top 10 At A Glance." *Computerworld* 1 December 1980: 11.
"Toward the 'Wired Society'." *World Business Weekly* 8 June 1981: 29–33.
"Trade Deficits Spur the Case for Protection." *Business Week* 2 November 1981: 48.
"Transborder Telecommunications." *Telephony* 2 November 1981: 13.
Treadwell, David. "U.S. Warns Japan it Risks Trade War," *Philadelphia Inquirer* 7 November 1981: 4B.
"Truman Plants an Antitrust Time Bomb," *Business Week* 25 September 1948: 19.
"Two Hotels to Resell Phone Service for Profit under New FCC Regulation." *Telephony* 4 May 1981: 105.
"2 Satellites Placed in Orbit by 10 W. European Nations." *Philadelphia Inquirer* 20 June 1981: 3A.
Tymshare Corporation. *10-K Report to the Securities and Exchange Commission*, 31 December 1979.
"UK: Private Networks to be Allowed." *Intermedia* September 1981: 5.
U.S. Civil Action No. 74-1698. In U.S. District Court for District Of Columbia, USA, *Plaintiff*, v. American Telephone And Telegraph Company; Western Electric Company, Inc.; and Bell Telephone Laboratories, Inc., *Defendants, Defendants Pretrial Brief* 10 December 1980.
U.S. Civil Action No. 74-1698. In the U. S. District Court For District of Columbia. United States of America. *Plaintiff, v.* American Telephone and Telegraph Company; Western Electric Company, Inc.; and Bell Telephone Laboratories, Inc., *Defendants' Third Statement of Contentions and Proof*, 3 volumes, 10 March 1980.
U.S. Congress. House of Representatives 95th Congress 2d Session. Hearings Before A Subcommittee On Communications Of The Committee On Interstate And Foreign Commerce, 9, 10, 14, 15, 16 August 1978, On H.R. 13015 *The Communications Act Of 1978*, Vol. II Part 2, Serial No. 95-196. Washington, DC: USGPO 1979.
U.S. Congress. House of Representatives 96th Congress 1st Session. Hearings Before The Subcommittee On Communications Of The Committee On Interstate And Foreign Commerce, 24–26 April, 1–4, 8 May 1979, On H.R. 3333 *The Communications Act Of 1979*, Titles I and III, General Provisions, Vol. I Part 1, Serial No. 96-121, Washington, DC: USGPO 1980.
U.S. Congress. House of Representatives 96th Congress 2d Session. Hearings Before The Subcommittee On Transportation, Aviation, And Communications, And The Subcommittee On Science, Research, And Technology Of The Committee On Science And Technology, 20-22, 28 May 1980, *Communications Research And Development*. Washington, DC: USGPO 1980.
U.S. Congress. House of Representatives 94th Congress 2d Session. Hearings Before The Subcommittee On Communications Of The House Committee On Interstate And Foreign Commerce, 28–30 September 1976, *Hearings On Competition In The Telecommunications Industry*, Serial No. 94-129. Washington, DC: USGPO: 1977.
U.S. Congress. House of Representatives 96th Congress 1st Session. Hearings Before The Task Force On Government Inefficiency Of The Committee On The Budget, 9 November 1979, *Competitive Procurement*. Washington, DC: USGPO 1980.
U.S. Congress, House of Representatives 97th Congress 1st session. *Congressional Record* 19 February 1981: H532.
U.S. Congress. House of Representatives 96th Congress 2d Session. Hearings Before The Subcommittee On Government Information And Individual Rights Of The Committee On Government Operations, 10, 13, 27 March and 21 April 1980, *Hearings On International Data Flow*. Washington, DC: USGPO 1980.
U.S. Congress. House of Representatives 94th Congress 1st Session. Subcommittee On Communications, Committee On Interstate And Foreign Commerce, December 1975, *Interim Report And Recommended Courses Of Action Resulting From The Hearings On Telecommunications Research And Policy Development*. Washington, DC: USGPO 1975.
U.S. Congress. House of Representatives 97th Congress 1st Session. Hearings Before The Subcommittee On Government Information And Individual Rights Of The Committee On Government

Operations, 31 March and 2 April 1981, On H.R. 1957, *International Communications Reorganization Act Of 1981*. Washington, DC: USGPO 1981.

U.S. Congress. House of Representatives 96th Congress 2d Session. Hearings Before A Subcommittee Of The Committee On Government Operations, 7, 21, 26 February 1980, On. H.R. 6410, *Paperwork Reduction Act Of 1980*. Washington, DC: USGPO 1980.

U.S. Congress. House of Representatives 96th Congress 2d Session. 19 March 1980, Report No. 96-835, *Paperwork Reduction Act Of 1980*.

U.S. Congress. House of Representatives 95th Congress 1st Session. Hearings Before A Subcommittee Of The Committee On Government Operations, 17 October 1977, *Privacy And Confidentiality Report And Final Recommendations Of The Commission On Federal Paperwork*. Washington, DC: USGPO 1977.

U.S. Congress. House of Representatives 97th Congress 1st Session. Committee On Foreign Affairs, May 1981, *Science, Technology, And American Diplomacy 1981*. Second Annual Report Submitted to the Congress by the President Pursuant to Section 503(b) of Title V of Public Law 94-426. Washington, DC: USGPO 1981.

U.S. Congress. House of Representatives 97th Congress 1st Session. Hearings Before The Subcomittee On Telecommunications, Consumer Protection, And Finance Of The Committee On Energy And Commerce, 20, 27, 28 May 1981, *Status Of Competition And Deregulation In The Telecommunications Industry*, Serial No. 97-29. Washington, DC: USGPO 1981.

U.S. Congress. House of Representatives 96th Congress 2d Session. Hearings Before The Subcommittee On Monopolies And Commercial Law Of The Committee On The Judiciary, 9, 16 September 1980, On H.R. 6121 *Telecommunications Act Of 1980*, Serial No. 69. Washington, DC: USGPO 1981.

U.S. Congress. Senate 95th Congress 1st Session. Hearings Before The Subcommittee On Manpower And Personnel Of The Committee On Armed Services, 12 July 1977, *Contracting Out Of Defense Functions And Services*. Washington, DC: USGPO 1977.

U.S. Congress. Senate 96th Congress 2d Session. Hearings Before The Committee On Armed Services, 5, 11–14, 25, 26 May 1980, On S. 2294 *Department Of Defense Authorization For Appropriations For Fiscal Year 1981*, Part 5 Research And Development. Washington, DC: USGPO 1980.

U.S. Congress. Senate 96th Congress 1st Session. Hearings Before The Subcommittee On Civil Services And General Services Of The Committee On Governmental Affairs. *Federal Government's Use Of Consultant Services*. Washington, DC: USGPO 1980.

U.S. Congress. Senate 95th Congress 1st Session. Hearings Before The Subcommittee On Communications Of The Committee On Commerce, Science, And Transportation, 21, 22 March 1977, *Hearings On Domestic Telecommunication Common Carrier Policies*, Two Parts, Serial No. 95-42. Washington, DC: USGPO: 1977.

U.S. Congress. Senate 93d Congress 1st Session. Hearings Before A Subcommittee On Antitrust And Monopoly Of The Committee On The Judiciary, 30, 31 July, 1, 2 August 1973, On S. 1167 *Hearings On The Industrial Reorganization Act*, Part Two The Communications Industry. Washington, DC: USGPO 1973.

U.S. Congress. Senate 95th Congress 2d Session. Staff Study For The Committee On Governmental Affairs. *Interlocking Directorates Among The Major U.S. Corporations*. Washington, DC: USGPO January 1978.

U.S. Congress. Senate 97th Congress 1st Session. Hearings Before The Subcommittee On Communications Of The Committee On Commerce, Science, And Transportation, 18 February 1981, On S. 271 *International Record Carrier Competition Act Of 1981*. Serial No. 97-5. Washington, DC: USGPO 1981.

U.S. Congress. Senate 95th Congress 1st Session. Hearings Before The Subcommittee On Communications Of The Committee On Commerce, Science, and Transportation, 13 July 1977, *International Telecommunications Policies*, Serial No. 95-54. Washington, DC: USGPO 1978.

U.S. Congress. Senate 96th Congress 1st Session. Hearings Before The Subcommittee On Federal Spending Practices And Open Government Of The Committee On Governmental Affairs, 1 November 1979, On S. 1411 *Paperwork And Redtape Reduction Act Of 1979*. Washington, DC: USGPO 1979.

U.S. Congress. Senate 96th Congress 2d Session. Hearings Before The Committee On Commerce, Science, And Transportation, 24, 25 September 1980, On S. 3003 +iService Industries Development Act, Serial No. 96-125. Washington, DC: USGPO 1981.

U.S. Congress. Senate 96th Congress 2d Session. Committee On Governmental Affairs. *Structure Of Corporate Concentration*, Staff Study, 2 volumes. Washington, D.C.: USGPO December 1980.

U.S. Congress. Senate 97th Congress 1st Session. Hearings Before The Subcommittee On Communications Of The Committee On Commerce, Science And Transportation, 2, 11, 15, 16, 19 June 1981, On S. 898 *The Telecommunications Competition And Deregulation Act Of 1981* Serial No. 97-61. Washington, DC: USGPO 1981.

U.S. Department Of Commerce, Bureau Of Industrial Economics. *1981 U.S. Industrial Outlook For 200 Industries With Projections For 1985*. Washington, DC: USGPO January 1981.

U.S. Department of Commerce, National Telecommunications And Information Administration. *1980 World's Submarine Cable Systems*. NTIA Contractor Deports, CR-80-6, May 1980.

U.S. Federal Communications Commission. *Annual Report* 1946–1979. Washington, DC: USGPO.

U.S. Federal Communications Commission. *Federal Communications Commission's List of Rules to be Reviewed Pursuant to Section 610 of the Regulatory Flexibility Act During 1981–1982*. General Docket No. 81-706, Notice Released 30 October 1981.

U.S. Federal Communications Commission. "Inquiry into Policy to be Followed in Future Authorization of Overseas Dataphone Service," FCC Docket No. 19558, (Notice of Inquiry Released 31 July 1972).

U.S. Federal Communications Commission. "In Re Petition Of Aeronautical Radio, Inc. For Declaratory Ruling Or Rulemaking That It Is An Authorized User Of The International Telecommunications Facilities Provided By The Communications Satellite Corporation Under The Communications Satellite Act Of 1962," FCC Docket No. 80-170 (Notice Of Proposed Rulemaking Released 6 May 1980).

U.S. Federal Communications Commission. "In the Matter of Allocation of Frequencies in the Bands Above 890 Mc," Docket No. 11866 (Preliminary Notice Of Hearing Released 9 November 1956).

U.S. Federal Communications Commission. "In the Matter of Amendment of Parts 2, 21, 74 and 94 of the Commission's Rules To Allocate Spectrum At 18 GHz," FCC Docket No. 79-188, RM-3247, RM-3497, (Further Notice Of Proposed Rulemaking Released 2 September 1981, Notice Of Proposed Rulemaking 1 August 1979.

U.S. Federal Communications Commission. "In the Matter of Amendment of Section 64.702 of the Commission's Rules and Regulations (Second Computer Inquiry)," FCC Docket No. 20828, (Notice Of Inquiry And Proposed Rulemaking 1976, Further Notice Of Proposed Rulemaking 17 May 1979).

U.S. Federal Communications Commission. "In the Matter of American Telephone and Telegraph Revisions to Tariff FCC No. 259 (WATS)," CC Docket No. 80-765, Transmittal No. 13555, 15 September 1980.

U.S. Federal Communications Commission. "In the Matter of Comsat Study—Implementation of Section 505 of the International Maritime Satellite Telecommunications Act," Docket No. 79-266, Final Report And Order Released 1 May 1980.

U.S. Federal Communications Commission. "In the Matter of Establishment of Policies and Procedures for Consideration of Applications to Provide Specialized Common Carrier Services in the Domestic Public Point-to-Point Microwave Service, and Proposed Amendments to Parts 21.43 and 61 of the Commission's Rules," FCC Docket No. 18920, (Notice Of Inquiry, Notice Of Proposed Rulemaking Released 17 July 1970).

U.S. Federal Communications Commission. "In the Matter of International Record Carriers' Scope of Operations in the Continental United States," FCC Docket No. 19660, RM-690, (Policy Statement and Order, 12 December 1979).

U.S. Federal Communications Commission. "In the Matter of Policy and Rules Concerning Rates for Competitive Common Carrier Services and Facilities Authorization Therefor," CC Docket No. 79-252, (Notice Of Proposed Rulemaking 27 September 1979).

U.S. Federal Communications Commission. "In the Matter of Policy to be Followed in Future Licensing of Facilities for Overseas Communications," FCC Docket No. 18875, (Further Notice Of Inquiry, 25 February 1975).

U.S. Federal Communications Commission. "In the Matter of Regulatory and Policy Problems Presented by the Interdependence of Computer and Communication Services and Facilities," FCC Docket No. 16979 (Notice of Inquiry 10 November 1966).

U.S. Federal Communications Commission. "In the Matter of Regulatory Policies Concerning Resale and Shared Use of Common Carrier Domestic Public Switched Network Services," CC Docket No. 80-54, RM-3453 (Notice of Proposed Rulemaking 19 February 1980).

U.S. Federal Communications Commission. "In the Matter of Regulatory Policies Concerning Resale and Shared Use of Common Carrier International Communications Services," CC Docket No. 80-176 (Notice of Proposed Rulemaking, 19 May 1980).

U.S. Federal Communications Commission. "In the Matter of Regulatory Policies Concerning Resale and Shared Use of Common Carrier Services and Facilities," FCC Docket No. 20097 (Report and Order, 16 July 1976).

U.S. Federal Communications Commission. "In the Matter of Revisions to Tariffs for Establishing Separate Charges for Terminals, Tielines, and Transmission Offered in Connection with International Telex Service and Implementing Expanded Gateways and Additional Domestic Operating Areas for International Telecommunications Service," FCC Docket No. 80-339 (Memorandum Opinion and Order, 8 August 1980).

U.S. Federal Communications Commission. "In the Matter of Use of the Carterfone Device in the Message Toll Telephone Service," Docket No. 16942 (Notice of Inquiry, 21 October 1966).

U.S. Federal Communications Commission. *Statistics Of Communications Common Carriers*, 1941, 1946, 1951, 1956, 1961, 1966, 1971, 1976, 1979. Washington, DC: USGPO.

U.S. *Federal Register*. "Postal Service," Vol. 46 No. 199 (15 October 1981), Notices: 50875.

U.S. General Accounting Office. PSAD-79-69, 18 May 1979 (untitled).

U.S. General Accounting Office. Report to the Congress by the Comptroller General of the United States. "Accounting For Automatic Data Processing Costs Needs Improvement." FGMSD-78-14, 7 February 1978.

U.S. General Accounting Office. Report to the Congress by the Comptroller General of the United States, "The Federal Information Processing Standards Program: Many Potential Benefits, Little Progress, And Many Problems." FGMSD-78-23, 19 April 1978.

U.S. General Accounting Office. Report by the Comptroller General of the United States, "Federal Paperwork: Its Impact On American Business." GGD-79-4, 17 November 1978.

U.S. General Accounting Office. Report by the Comptroller General of the United States, "Implications Of Electronic Mail For The Postal Service's Work Force." GGD-81-30, 6 February 1981.

U.S. General Accounting Office. Report to the Director, Office of Management and Budget, "Independent Regulatory Agencies Can Reduce Paperwork Burden On Industry." AFMD-81-70, 7 July 1981.

U.S. General Accounting Office. Report by the Comptroller General of the United States, "Shifting The Government's Automatic Data Processing Requirements To The Private Sector: Further Study And Better Guidance Needed." FGMSD-78–22, 11 April 1978.

U.S. General Accounting Office. Report to the Congress by the Comptroller General of the United States, "Wider Use Of Better Computer Software Technology Can Improve Management Control And Reduce Costs." FSMSD-80-30, 29 April 1980.

U.S. General Services Administration, Automated Data And Telecommunications Service. "Automatic Data Processing Activities In The United States Government as of the end of Fiscal Year 1979, Summary." Washington, DC: USGPO, September 1980.

U.S. National Science Foundation. *Research And Development In Industry 1977*. Washington, DC: USGPO 1977.

U.S. Office of the United States Trade Representative. "A Preliminary Review Of Barriers To Trade In Telecommunications, Data Processing, And Information Services And Transborder Data Flows; Motivations For Imposing Barriers, Trade Implications And Possible Approaches To Resolution," 3 September 1980 (Draft).

U.S. President, Council Of Economic Advisors. *Economic Report Of The President*, Transmitted to the Congress January 1981. Washington, DC: USGPO 1981.

U.S. President's Task Force On Communications Policy. *Final Report*. Washington, D.C.: USGPO 7 December 1968.

United States Of America, v. International Business Machines Corporation, *Pretrial Brief For The United States*, 69 Civ. 200 (DNE) United States District Court Southern District Of New York, 17 October 1974.

United States of America, v. Western Electric Company, Inc., and American Telephone and Telegraph Company, 13 RR 2143; 1956 Trade Case 71,134 filed 24 January 1956.

"U.S. Trading Partners Are Worried On Use of Barriers and Subsidized Export Loans," *Wall Street Journal* 15 June 1981: 6.

"USA: the Ambitions of SBS." *Intermedia* September 1981: 3.

User Group Survey—Deutsche Telecom eV, West Germany," *Telecommunications Policy* Vol. 4 No. 3 (1979): 228–229.

"User Group Survey—Telecommunications Manager Association, Belgium," *Telecommunications Policy* Vol. 4 No. 1 (1979): 63.

"Users, Unions Call for More Investment in British Telecom." *Telephony* 23 March 1981: 38-39.

Uttal, Bro. "How to Deregulate AT&T," *Fortune* 30 November 1981: 70–75.

Van Zandt, Howard F. "East Opens Its Doors to the West," *Telephony* 1 June 1981: 26-27.

"Wall Street Group Elects New Officers," *Communications News* September 1981: 15.

Wallerstein, Immanuel. "Friends as Foes," *Foreign Policy* 40 (1980): 119–131.

"WASEC Keeps Perking Along, Plans To Add 3 More Channels," *Variety* 2 September 1981: 38.

Wessler, Barry D. "United States Public Packet Networks: an Update," *Telecommunication Journal* Vol. 47 No. VI (1980): 373–375.

"When TI Talks, the Message Moves Fast." *Electronics* 17 January 1980: 103–104.

Whichard, Obie G. "U.S. Direct Foreign Investment Abroad in 1979," U.S. Department of Commerce, Bureau of Economic Analysis *Survey Of Current Business* Vol. 60 No. 8 (1980): 16–36.

Wiley, Richard E. "Competition and Deregulation in Telecommunications: The American Experience," Chapter 2, 37–59 in Leonard Lewin, ed., *Telecommunications in the United States: Trends and Policies*. Dedham, MA: Artech House 1981.

Wilkins, Mira. *The Maturing Of Multinational Enterprise*. Cambridge MA: Harvard University Press 1974.

Williams, William Appleman. "Thoughts On Rereading Henry Adams." *Journal of American History*, June 1981 (Vol. 68, No. 1): 7–15.

"Yankee Group Sees Changes in AT&T Goals." *Communications News*, September 1981: 18.

Zahn, Paul. "EEC Plays Hardball with Japanese on Exports," *Advertising Age* 10 November 1980: 68.

Index

A

"Above 890" decision, 8–15, 73
Ad Hoc Telecommunications Users Committee, 76, 86–87
Advanced Communications Service, 82, 89
Advertising, foreign, 113–114
Advertising Age, 1
Advisory Committee for Information Science and Technology, 140
Advisory Committee on International Communications and Information, 141, 142
Aeronautical Radio, Inc. (ARINC), 23, 72, 74
Aerospace Industries Association of America (AIA), 23, 26, 38, 76
Aetna Life and Casualty Company, 23, 28, 31, 36, 49, 90, 175
Ahern, V. M., 103
Airlines, 23
Allied Chemical Corporation, 153
Amax, 43
American Bankers Association (ABA), 23, 27–28, 31, 34, 38, 40, 45–46, 62
American Business Press, Inc., 23
American Enterprise Institute, 14
American Express Company, 76, 81, 83, 86, 205–206
American Newspaper Publishers Association (ANPA), 9, 23, 72
American Petroleum Institute (API), 13, 17, 18, 23, 25–26, 29, 31, 34, 38, 39, 40, 44–45, 62, 72, 76, 84, 154
American Satellite Corporation, 49, 50, 51, 175
American Telephone & Telegraph Company (AT&T), 5, 6, 213
 "Above 890" decision, 9, 13–14, 61
 Advanced Communications Service, 82, 89
 Annual Report (1980), 15, 165
 antitrust suits against, 70, 71, 82, 91–93, 194
 Carterfone decision, 16, 21, 61
 Dataphone service, 152–155
 Domestic Satellite decision, 49, 61
 First Computer Inquiry, 34–35, 38–41
 International, 161, 168
 legislative battle for position, 54–61

AT&T *(cont.)*
 resale and sharing issue, 72, 73, 74, 76, 77, 84
 rivalry with Western Union, 150
 Second Computer Inquiry, 85–89, 164, 168
 Specialized Common Carrier proceeding, 42, 44, 46–48, 61
 TAT 4 decision, 151
American Trucking Associations, Inc., 23
Anderson, H., 54
Antilles, 112
Arbitron, 82
Argentina, 214
Arianespace, 178
Armco Steel Inc., 27
Associated Press (AP), 49, 73
Association of American Railroads (AARR), 23, 24, 31, 34, 36, 38, 85
Association of Data Communications Users, 57, 59, 62
Association of Data Processing Service Organizations (ADAPSO), 37, 79, 90
Association Francaise des Utilisateurs du Telephone et des Telecommunications (AFUTT), 114–115
Association Suisse des Usagers de Telecommunications (ASUT), 110–111, 117
Atlantic Richfield (ARCO), 74
Automobile Manufacturers Association, 10–11
Aviation Week & Space Technology, 193
Avis Rent A Car System, 76–77

B

"Baby Bell", 164–165
Bache Company, 83
Baker, K., 119
Bandwidth, 5, 6, 32, 52, 151
Bank of America, 78
Bank Wire, 73
Barclays Merchant Bank, 177
Bargellini, P. L., 175
Barna, B., 81
Barnwell, R., 63, 64, 99
Bartlett, J., 159
Belgium, 110, 185, 214
Bell Canada, 121

229

Bell Communications Systems, 121
Bell Laboratories, Inc., 91, 92, 149, 165, 167
Bell System, *see* American Telephone & Telegraph Company (AT&T)
Bethlehem Steel Corporation, 43, 86
Betteridge, W. W., 104, 162
Bird, F. S., 17–18
Block, V., 75–76, 93, 121
Blumenthal, M., 80, 81, 94
Boeing Computer Services, 26, 80, 153–154
Bolger, W. F., 213, 215
Bolter, W. G., 95
Booz, Allen and Hamilton Inc., 22, 23, 24, 32
Borghesani, W. H., 23, 42
Brady, R. A., 63
British Aerospace, 177
British Petroleum, 177
British Post Office, 173, 174
British Telecom International, 118, 119–121, 175, 176
British Telecommunications Act, 119, 121
Brock, G., 5, 7, 8, 13, 14, 21, 42, 43, 47, 48, 49, 74, 85, 191, 195, 196, 200
Brown, C. L., 1, 97, 167, 195
Brown, G. R., 144
Bruce, R. R., 97, 104, 109, 120, 147–148, 174, 180, 182, 206
Buckley, J. L., 142–143
Bundled prices, 157–161
Bureau of Labor Statistics (BLS), 209
Burkert, H., 105, 116
Burroughs Corporation, 81, 82
Burtz, L., 185
Business Advisory Council on Federal Reports, 205
Business Equipment Manufacturers Association *see* BEAMA (CBEMA), 22, 24, 32, 37, 38
Business Week, 21, 79, 161, 184

C

Cable and Wireless, Inc., 177
Caldwell, J. E., 9–10
Camrass, R., 119
Canada, 112, 118, 121, 175, 176, 214
Capone, R., 66
Carr, F. J., 204
Carterfone decision, 15–22, 30, 32, 47, 55, 61, 87, 104
Chace, S., 82, 149
Chamoux, J. P., 114, 118
Chandler, A. D., 26
Chase Manhattan Bank, 81, 100–101, 109, 187
Chevron Oil, California Company Division, 17
Chrysler Corporation, 43, 86
Chung, W. K., 112, 118

Citicorp, 73, 77, 83, 186
Citishare, 80, 83
Civil Aeronautics Board (CAB), 209
Clyne Report, 122
COBOL (Common Business-Oriented Language), 196
Colby, W. E., 143–144, 176
Coldwell, Banker & Company, 83
Cole, R. J., 83
Colgate-Palmolive Company, 113
Commerce, U. S. Department of, 3, 4, 99
 1980 World's Submarine Telephone Cable Systems, 105
 1981 U. S. Industrial Outlook for 200 Industries With Projections for 1985, 112, 165, 192
Committee of Corporate Telephone Users (CCTU), 77–78
Communications Act of 1934, 37, 38, 63, 66, 86, 88, 150, 155, 169, 172
Communications Act of 1978, 67–69
Communications Act of 1979, 155, 180, 182
Communications News, 51, 62
Communications Regulatory Commission (CRC), 68, 69
Competition, 7, 66, 67, 105, 107, 150
Competitive Common Carrier Proceeding, 84–86
Comp-U-Card, 51
Computer and Business Equipment Manufacturers Association (CBEMA), 154
Computer Sciences Corporation, 198–199
Computers, estimated distribution of installations by user industry, 24
Computerworld, 54
Comsat, 48–49, 90, 169–172, 175, 177–178
 Annual Report (1980), 50
Comsat General, 177
Congress of the United States, *see* House of Representatives hearings; Senate hearings
Congressional Record, 140, 148, 168
Consumer Communications Reform Act of 1976 (Bell Bill), 56, 57, 61, 66, 70
Consumer credit, 4–5
Continental Illinois National Bank and Trust Company, 136–137
Control Data Corporation, 76, 82, 90, 186, 198, 200
Corporate Information Center, 100
Corporate structure, 27
Cost-based pricing, 73, 147, 160, 161
Council of International Communications and Information, 141–142
Courtney, J., 43, 44
Cox, K., 195
Craig, L. C., 100
Crane, R. J., 109

Index   231

Credit Data Corporation, 23
Criner, J. G., 73
Crocker Bank, 51
Cybernet, 82

D

Darlington, R., 120
Data General Corporation, 200
Data processing companies
　government market (1979), 197
　top 100, 193
DataComm User Incorporated, 56
Datamation, 83, 197–199
Dataphone service, 152–155
Davignon plan, 119
De Jonquieres, G., 177
Dean Witter Reynolds, Inc., 83
deButts, John D., 1, 61, 94–95
Deere, John, Company, 110
Defense, U. S. Department of, 53, 151, 169, 174, 191–195
Defense Acquisition Regulation 9-602, 198
Delauer, R. D., 193–195
Deutsche Bundespost, 116
Deutsche Telecom eV, 110, 114, 115, 116
Digital Equipment Corporation, 197, 200
Direct foreign investment (U.S.), 99, 101, 108, 111–112, 118
Direction Des Affaires Industrielles Et Internationales, Direction Generale Des Telecommunications, *Annual Report* (1980), 115
Domestic Satellite decision, 48–52, 61
Donne, M., 177
Dow Chemical, Inc., 27
Dow Jones & Company, Inc., 49, 52, 73, 84–85, 178–179
Dun & Bradstreet, Inc., 81
Dunn, W. L., 178–179
Dupont de Nemours, E. I. & Company, 43, 62, 76, 86

E

Eastern Airlines, Inc., 23
Eaton Corporation, 153
Economic growth, 3–4
Economic Report Of The President (1981), 4, 10
Economist, The, 166, 167
EDP Industry Report (1980), 6
Eger, J. M., 104, 109, 137, 183
Electronic Computer Originated Mail (E-COM), 212–213
Electronic Data Systems (EDS), 198

Electronic Industries Association (EIA), 27, 28, 31–32, 34, 36, 37, 38, 46, 154
Emmett, R., 83
Energy, Department of, 203
Energy Consumption, 3–4
Energy and Environment Analysis, 203
Energy Research and Development Administration, 203
English, G., 140, 143, 213
European Conference of Postal and Telecommunications (CEPT), 119, 171, 181
EUSIDIC, 121
Evans Economic Data Base, 82
Exxon Corporation, 86

F

Fairchild Industries, 192
Fargo, D. S., 101, 106
Federal Communications Commission (FCC)
　Annual Report (1946–1979), 4, 16, 152, 154
　on Consumer Communication Reform Act, 70–71
　Docket No. 81-706, 209
　Docket No. 19558, 152–155
　Docket No. 80-170, 178–179
　Docket No. 11866 ("Above 890"), 8–15, 73
　Docket No. 79-188, 94
　Docket No. 20828 (Second Computer Inquiry), 85–89, 164, 168
　Docket No. 80-765, 77
　Docket No. 79-266, 171, 178
　Docket No. 18920 (Specialized Common Carrier Proceeding), 41–48, 61, 104
　Docket No. 19660, 159, 181
　Docket No. 79-252 (Competitive Common Carrier Proceeding), 84–86
　Docket No. 18875, 171, 174
　Docket No. 16979 (First Computer Inquiry), 22–42, 53, 85, 104
　Docket No. 80-54, 76
　Docket No. 80-176, 63, 72–76, 99, 185, 186
　Docket No. 20097, 184
　Docket No. 80-339, 160
　Docket No. 16942 (Carterfone), 15–22, 30, 32, 47, 55, 61, 87, 104
　Docket No. 80-632, 155
　Domestic Satellite decision (1972), 48–52, 61
　Office of Plans and Policy, 138
　Statistics of Communications Common Carriers (1941–1979), 4, 5, 14, 101, 102, 157
　TAT 4 decision, 151, 155
　TAT 5 decision, 172–173
　TAT 7 decision, 173–175

Federal Energy Regulatory Commission, 209
Federal Express Corporation, 212
Federal Information Locator System, 207
Federal Information Processing Standards Coordinating and Advisory Committee, 140
Federal Maritime Commission, 209
Federal Register, 212
Federal Reports Act of 1942, 207
Federal Reserve System, 28, 109
Feketekuty, G., 125, 136, 143, 145–146
Ferris, C. D., 88, 159, 160, 167, 168, 171, 173–174, 178
Fettig, L., 200, 201
Fiberboard Corporation, 63
Firestone Tire & Rubber Company, 78, 81
First Computer Inquiry, 22–42, 53, 85, 104
First Data Resources, 81
First Interstate Bancorp, 50
Ford Aerospace, 192
Ford Motor Company, 76, 86, 100
Fortune magazine, 34
Fowler, M., 92
Fouch, G. G., 112, 118
France, 111, 112, 114–115, 178, 214
Frank, H., 54
Freeman, H. L., 99, 107, 123
Frequency Management Advisory Council, 140

G

Gallagher, E. A., 173
Garrett, A., 113–114
Geller, H., 103–104, 109, 123
General Accounting Office (GAO), 191, 196, 198, 199, 203, 205–207, 209, 210
General Agreements on Trade and Tariffs (GATT), 148
General Dynamics, 26, 50
General Electric Company, 7, 24, 26, 62, 76
General Motors Corporation (GM), 26, 62, 175
General Services Administration (GSA), 198, 199
  Automated Data and Telecommunications Service, 196, 197, 200, 201, 202
General Telephone & Electronics Corporation (GTE), 35, 53, 54, 82, 88, 90, 161, 170, 193
General Telephone and Electronics Satellite Corporation, 50
German Monopoly Commission, 116–117
Gilchrist, B., 191
Goldfarb, D. C., 150, 169
Goldstine, H. H., 5
Goldwater, B., 173
Grad, F. P., 150, 169
Granquist, W., 204

Graphnet, 72, 180, 181, 210, 213
Great Britain, 111, 112, 118–121, 171, 176–178, 185, 214
Greyhound, 77
Grove, D., 89
Grumman Data Systems, 26, 198
Guttman, D., 203

H

Hanrahan, J., 203
Hatfield, D., 145
Henkel Corporation, 113
Herman, E. S., 26, 34, 60, 63, 111, 139–140
Hewlett-Packard, 197
Hilsman, W. J., 193, 194
Hirsch, P., 70, 94, 157, 194, 213, 215
Hoard, B., 54, 109
Hodgson, J., 171, 173–174
Holiday Inn hotel system, 51
Holsendolph, E., 94
Honeywell, Inc., 5, 26, 198
Horton, F., 204, 205, 207
House of Representatives bills
  4758, 213
  5158, 89, 90
  8443, 140
House of Representatives hearings
  Communications Act of 1978 (1979), 67–69
  Communications Act of 1979 (1980), 155, 180, 182
  Communications Research And Development (1980), 206, 210–211, 214
  Competitive Procurement (1980), 200
  Hearings On Competition In The Telecommunications Industry (1977), 15, 56, 95
  Hearings On International Data Flow (1980), 101, 103, 104, 112, 123, 137, 144, 176–177, 183
  Interim Report And Recommended Courses Of Action Resulting From The Hearings On Telecommunications Research And Policy Development (1975), 139, 164
  International Communications Reorganization Act of 1981 (1981), 99, 104, 107, 109–110, 119, 120, 123, 125, 136, 142–143, 146, 147, 174
  Paperwork Reduction Act of 1980 (1980), 204, 206–208
  Privacy And Confidentiality Report And Final Recommendations Of The Commission On Federal Paperwork (1977), 204, 205, 207
  Science, Technology, And American Diplomacy 1981 (1981), 171

Status of Competition And Deregulation In The Telecommunications Industry (1981), 79, 80, 83, 104, 162
Telecommunications Act of 1980 (1981), 69
Hughes Corporation, 26, 49, 50
Humble Oil & Refining Corporation, 23
Hush-A-Phone, 16, 31, 75

I

ICL, 118
Independent Data Communications Manufacturers Association, 90
Industrial Radio Services, 18
  land fixed stations in (1950–1975), 14
Industry Sector Advisory Committee on Communication Equipment and Non-Consumer Electronic Equipment for Multilateral Trade Negotiations, 140
INFONET, 198
Information Science and Technology Act of 1981, 143–145
Inglis, A. F., 50
Institute for Information Policy and Research, 144
INTELPOST (International Electronic Post), 214, 215
Intelsat, 118, 169–171, 173, 175–178
Interconnection issue, 15–22, 30–32, 35
Interdepartmental Radio Advisory Committee, 138
Intermedia, 175
International Business Machines Corporation (IBM), 5, 7, 26, 29, 30, 49, 90, 108–109, 166, 168, 175, 177, 178, 197, 198, 200, 213
International Business Machines Information Network, 82
International Communications Association (ICA), 57, 62–63, 67–70, 90, 93, 99, 186–187
International Communications Reorganization Act of 1981, 99, 104, 107, 109–110, 119, 120, 123, 125, 136, 141–143, 146, 147, 174
International Data Corporation Census (1966), 24
International Electronic Facsimile Users Association, 61–62
International Maritime Satellite (Inmarsat), 178
International Record Carrier Competition Act of 1981, 160, 161, 173, 181
International Telecommunications Union (ITU), 108, 185
  Consultative Committee on Telephones and Telegraphs (CCITT), 121, 185–187

International Telecommunications Satellite Organization (Intelsat), 118, 169–171, 173, 175–178
International Telecommunications Users Group (INTUG), 121
International Telephone & Telegraph Corporation (ITT), 26, 90, 150, 157, 170, 172, 184, 213
Interstate Commerce Commission (ICC), 209
Iowa Telecommunications User Group, 62
Ireland, 121
Irwin, D. A., 95
ISA Communications Services, 175
Italy, 185

J

Jansky, D., 192
Japan, 107, 112, 148
Justice, U. S. Department of, 40, 82, 91–93, 213

K

Kaplan, G., 175
Karten, H., 7
Keatley, R., 107
Keith, D. R., 192
King, J. L., 191, 196
Kirchner, J., 142–145, 196, 206
K-Mart, 76
Knapp, G. F., 173
Kodak, 192
Kraemer, K. L., 191, 196
Kuwait, 185–186

L

Lee, R. E., 160, 182–183
Lee, S., 215
Lewis, P., 118
Litke, R. M., 94
Little, Arthur D., Inc., 114, 162
Local Digital Distribution Company, 51
Lockheed Aircraft Corporation, 23, 26
Lockheed Information Systems, 54

M

Malik R., 148
Manufacturers Hanover Trust Company, 54, 109
Marable, W. T., 210, 214
Maritime, Industrial and Aviation Radio Services, 19
Marks, H. E., 44
Martin, J., 7

Master Charge, 77, 78
McClintick, D., 211
McDonnell Douglas Automation, 80
McDonnell Douglas Corporation, 26
McGinley, L., 209
McGraw-Hill, Inc., 23
McGuire, T. J., 93
MCI Communications Corporation, 35, 42–43, 52, 66, 150
McNamar, R. T., 145
Melody, W. H., 15, 16, 74, 75
Midwestern Telecommunications Users, 62
Mitterand, F., 115
Mobil Corporation, 44
Monsanto Company, Inc., 9–10, 23, 43, 76, 86
Montgomery Ward & Company, 55
Moore, V. P., Jr., 64, 65
Morgan, J., 118–119, 123
Morgan, W. L., 49, 50
MTS, 75, 78, 84, 184
Murphy, E. F., 155, 173
Muzak, 49

N

National Aeronautics and Space Administration (NASA), 198, 202–203
National Association of Manufacturers (NAM), 11–12, 23, 27, 31, 34, 36, 38, 39, 40, 44, 45, 73
National Association of Regulatory Utility Commissioners, 93
National Bureau of Standards, 199
National Commission on Libraries and Information Science, 206
National Committee for Utilities Radio, 23
National CSS, Inc., 81
National Information Science and Technology Board, 144
National Library of Medicine, 54
National Retail Dry Goods Association (NRDGA), 10
National Retail Merchants Association (NRMA), 20, 23, 30, 33–34, 38, 39, 42, 66, 69, 73, 90
National Science Foundation, 144, 191
National Telecommunications and Information Administration (NTIA), 138, 144, 145, 182, 187, 213
NCR Corporation, 5
Netherlands, 112, 114, 118
New York Stock Exchange, 65
New York Times, 51
Nichols, R. B., 151
Nippon Telegraph and Telephone Public Corporation, 148–149
Nora/Minc Report, 122

O

Oettinger, A., 195
Office of Information and Regulatory Affairs, 207
Office of Management and Budget (OMB), 200–205, 207–209
Office of Naval Research, 24
Office of Telecommunications (Commerce Department), 138, 139
Office of Telecommunications Policy (Executive Office of the President), 138, 139
Office of the U. S. Trade Representative, 144, 145
Olin Corporation, 76, 86
O'Neill, P., 110, 115, 120, 121
Onstad, P., 186
"Open skies" satellite policy, 138, 179
Organization for Economic Cooperation and Development (OECD), Business and Industry Advisory Committee to, 121–122
O'Rourke, T. J., 79, 80, 83
Overseas Dataphone decision, 152–155

P

Packet Communications Inc. (PCI), 53, 54, 72
Packet switched networks, 52–54, 64, 108, 183
Paperwork and Redtape Reduction Act of 1979, 205
Paperwork Reduction Act of 1980, 203–204, 206–210
Pelka, F. M., 44
Peltu, M., 119, 121
Penney, J. C., Company, Inc., 23, 66, 76, 86
Petroleum industry, 13, 17–20, 25–26
Petroleum Radio Service, 13
Phibro Corporation, 83
Pierre, P. A., 192
Pine, A., 145
Pomeroy, W. B., 44
Postal Rate Commission, 213, 214
Postal service, 210–215
Postal Telegraph, 150
Posts, Telephones and Telegraphs (PTTs), 103, 105, 113–117, 119–122, 123, 147, 169, 170–174, 176, 177, 179, 181–186, 214
President's Project on Information Technology and Government Reorganization, 207
President's Task Force on Communications Policy, *Final Report* (1968), 14, 41, 43, 138
Preyer, R., 101, 140, 176
Price-Wen, K., 73n
Private Mail Carriage Act of 1981, 215
Private Radio Services, 13, 14, 18, 85

Index

Procter and Gamble Company, 113
Protectionism, 107, 145, 148
Prudential Insurance Company of America, 83
Public Law 89-306 (Brooks Act), 199

R

Radio Liberia, 81
Railroad industry, 24–25, 30–31, 85
Ransome, R., 18–19
RCA Americom, 50, 51
RCA Corporation, 26, 49, 50, 157, 170, 172, 184, 192
Reciprocity, 147–149
Reis, V. G., 11–12
Remington Rand, 5
Republic Steel Corporation, 86
Resale and sharing issue, 36–37, 104, 183–187
Reynolds & Reynolds Company, 76, 86
Richardson-Merrell Company, 113
Riddle, K., 101
Rubiner, B., 118

S

Sageman, R. E., 165
Salim, M., 117, 118
Salomon Brothers, 83
Satellite Business Systems (SBS), 49, 50, 51, 90, 175–178, 210
Satellites, 64, 159, 169–179
Saunders, G., 91
Saxton, W. A., 56–59
Scannell, T., 34, 202
Schiller, H. I., 169, 170, 171
Schultz, B., 82, 109, 118
Sears Roebuck and Company, 23, 62, 76, 83, 86
Second Computer Inquiry, 85–89, 104, 164, 168
Securities and Exchange Commission (SEC), 197
Securities Industry Automation Corporation (SIAC), 64, 65, 66, 73, 179
Securities Telecommunications Organization (SECTOR), 64, 65
Senate hearings
 Contracting Out of Defense Functions and Service (1977), 200, 201
 Department of Defense Authorization For Appropriations For Fiscal Year 1981 (1980), 192
 Federal Government's Use Of Consultant Services (1980), 203
 Hearings On Domestic Telecommunication Common Carrier Policies (1977), 41, 48, 53, 63–66, 70, 99

235

 Hearings On The Industrial Reorganization Act (1973), 55
 Interlocking Directorates Among The Major U. S. Corporations (1978), 60
 International Record Carrier Competition Act Of 1981 (1981), 160, 161, 173, 181
 International Tele-Communications Policies (1978), 151, 156–158, 169–173, 180
 Paperwork And Redtape Reduction Act Of 1979 (1979), 205
 Service Industries Development Act (1981), 145
 Structure Of Corporate Concentration (1980), 60
 Telecommunications Competition and Deregulation Act of 1981, 88, 90
Service Industries Development Act, 145
Shared use issue, 36–37, 104, 183–187
Shayon, R. L., 154
Shearson Loeb Rhoades, Inc., 83
Sheley, W. H., Jr., 199
Shell Oil Company, 86
Shell Communications, 18–19
Sherman, K., 53
Singer Company, 23
Sirbu, M. A., Jr., 108
Slater, C., 209
Smith, A., 168
Societe Generale, 112
Societe Internationale de Telecommunications Aeronautique, 23
Sodolski, J., 46
Sorkin, A. L., 210, 211, 213, 214
Southern Pacific Communications (SPC), 35, 49, 50, 66, 81
Southern Pacific Railroad, 81
Space Systems and Technology Advisory Committee, 140
Space and Terrestrial Applications Advisory Committee, 140
Special and Safety Radio Services, 13, 18
Specialized Common Carrier Proceeding, 41–48, 61, 104
Sperry Rand, 26, 198
Sperry Univac, 5, 49, 200
Staats, E. B., 201
Standard & Poor's Industry Surveys, *Office Equipment Systems and Services* (1980), 24, 195
State, U. S. Department of, 142, 143, 182
Stolarow, J. H., 198
Strassman, P. A., 206
Stritzler, W. P., 163
Stubblebine, A. N., III, 192
Submarine cables, 105, 151
Sullivan, D. J., Jr., 164
Sun Information Systems, 80

Supreme Court of the United States, 30
Switzerland, 110–111, 112, 117–118, 214
System Development Corporation, 54, 81, 82

T

TAT 4 decision, 151, 155
TAT 5 decision, 172–173
TAT 7 decision, 173–175
Tauke, T. J., 93
Tele-Cause, 90
Telecommunications Act of 1980, 69
Tele-Communications Association (TCA), 62, 63, 66, 90
Telecommunications Competition and Deregulation Act of 1981, 88, 90, 148, 168, 195
Telecommunications Managers Association, U. K., (TMA), 111, 119–120
Telecommunications Managers Association, Belgium, (TMAB), 110
Telecommunications Technology Research Advisory Council, 139
Telegraph, 150, 156–157
Telenet, 53, 54, 72, 82, 180–182
Telephone construction expenditures, international, 106
Telex, 101, 102, 156–161
Telepak, 13–14, 36, 43, 46, 47, 73–74
Terminal equipment markets, 21, 31, 35
Texas Instruments (TI), 100, 159
Thomas, C., 52
Thomas, W., 52
Thompson, G., 122
Ticketron, 82
Time Inc., 77
Times-Mirror Satellite Entertainment, 51
Timesharing, 37, 53
Topol, S., 51
Toth, V. J., 94
Trade barriers, 123–135
Travellers Insurance, 175
Treasury, U. S. Department of the, 211
Tropical Radio Telegraph, 150
TRT Telecommunications, 81, 150, 172, 185
Tymnet, 51, 53, 79
Tymshare Corporation, 78–79, 80

U

UHF, 18
Unilever Inc., 113
Union Carbide Corporation, 12, 43, 86
Union Pacific Railroad, 23
United Airlines, Inc., 23
United Fruit Company, 81, 150
United Information Systems, 80, 82

United Kingdom, 111, 112, 118–121, 171, 176–178, 185, 214
United Parcel Service (UPS), 211
United Press International (UPI), 49
United States National Committee for the International Consultative Committee on Telephones and Telegraphs of the ITU, 140
United States National Committee for the International Radio Consultative Committee of the ITU, 140
United States Postal Service (USPS), 210–215
United States Steel Corporation, 62, 76, 86
United States v. American Telephone and Telegraph Company; Western Electric Company, Inc., and Bell Telephone Laboratories, Inc. (1980), 91, 166, 167
United States v. Western Electric Company, Inc., and American Telephone and Telegraph Company (1956), 40, 91, 92
United Telecommunications, 82
User organization, 61–63 *see also,* individual trade associations
Utilities Telecommunications Council, 44
Uttal, B., 91

V

Value added networks (VANs), 52–54, 180–183
Van Deerlin, L., 57–59
Van Zandt, H. F., 149
VHF, 17, 18
VISA, 76, 77, 78

W

Walker, P. M., 161, 181
Walker, R. E. L., 112, 136–137
Wall Street Journal, 49, 50, 51, 166, 178, 209, 211
Wallerstein, I., 107
Warkow, C. W., 55
Warner Amex Satellite Entertainment, 51
Washburn, A., 174–175
WATS, 75–78, 84, 184, 212
Wells Fargo, 175
Wessel, M. R., 191
Wessler, B. D., 53
West Germany, 110, 112, 114, 115–117, 178, 214
Western Electric Company, Inc., 91, 149, 161–162, 164, 165, 167
Western Union, 49, 50, 72, 150, 151, 156–160
Western Union Cables, 150, 156
Western Union International (WUI), 150, 156, 157, 172, 185

Westinghouse Electric Corporation, 26, 27, 52, 76, 86, 175, 192
Weyerhauser Company, 27, 43
Whichard, O. G., 99
White, J. A., 82
Wiley, R. E., 14, 15, 41, 48, 154–158, 169–173, 180
Wilkins, R. E., 99
Williams C., 9
Williams, W. A., 189
Woody, M., 63, 70
World Business Weekly, 106

X

X.25 standard, 108–109
Xerox Corporation, 150, 152–153

Y

Yellow Pages, 92

Z

Zahn, P., 107